MECHANICAL DRAWING PROBLEMS

BY

CHARLES WILLIAM WEICK, B. Sc.

ASSISTANT PROFESSOR OF DRAWING AND DESIGN, TEACHERS COLLEGE
COLUMBIA UNIVERSITY, IN THE CITY OF NEW YORK; AUTHOR
OF "ELEMENTARY MECHANICAL DRAWING."

FIRST EDITION

British Library Cataloguing-in-Publication Data
A catalogue record for this book is available from the
British Library

Technical Drawing and Drafting

Technical drawing, also known as 'drafting' or 'draughting', is the act and discipline of composing plans that visually communicate how something functions or is to be constructed.

It is essential for communicating ideas in industry, architecture and engineering. The need for precise communication in the preparation of a functional document distinguishes technical drawing from the expressive drawing of the visual arts. Whereas artistic drawings are subjectively interpreted, with multiply determined meanings, technical drawings generally have only one intended meaning. To make the drawings easier to understand, practitioners use familiar symbols, perspectives, units of measurement, notation systems, visual styles, and page layout. Together, such conventions constitute a visual language, and help to ensure that the drawing is unambiguous and relatively easy to understand.

There are many methods of constructing a technical drawing, and most simple among them is a sketch. A sketch is a quickly executed, freehand drawing that is not intended as a finished work. In general, sketching is a quick way to record an idea for later use, and architects sketches in particular (in a very similar manner to fine artists) serve as a way to try out different ideas and establish a composition before undertaking more finished work. Architects drawings can also be used to convince clients of the merits of a design, to enable a building constructer to use them, and as a record of completed work. In a similar manner to engineering (and all other technical drawings), there is a set of conventions (i.e particular views, measurements, scales, and cross-referencing) that are utilised.

As opposed to free-sketching, technical drawings usually utilise various manuals and instruments. The basic drafting procedure is to place a piece of paper (or other material) on a smooth surface with right-angle corners and straight sides – typically a drawing board. A sliding straightedge known as a 'T-square' is then placed on one of the sides, allowing it to be slid across the side of the table, and over the surface of the paper. Parallel lines can be drawn simply by moving the T-square and running a pencil along the edge, as well as holding devices such as set squares or triangles. Other tools can be used to draw curves and circles, and primary among these are the compasses, used for drawing simple arcs and circles. Drafting templates are also utilised in cases where the drafter has to create recurring objects in a drawing – a massive time-saving development.

This basic drafting system requires an accurate table and constant attention to the positioning of the tools. A common error is to allow the triangles to push the top of the T-square down slightly, thereby throwing off all the angles. Even tasks as simple as drawing two angled lines meeting at a point require a number of moves of the T-square and triangles, and in general drafting this can be a time consuming process. In addition to the mastery of the mechanics of drawing lines, arcs, circles (and text) onto a piece of paper – the drafting effort requires a thorough understanding of geometry, trigonometry and spatial comprehension. In all cases, it demands precision and accuracy, and attention to detail.

Conventionally, drawings were made in ink on paper or a similar material, and any copies required had to be laboriously made by hand. The twentieth century saw a shift to drawing on tracing paper, so that mechanical copies could be run off efficiently. This was a substantial development in the drafting process – only eclipsed in the twenty-first century with 'computer-aided-drawing' systems (CAD). Although classical draftsmen and women are still in high demand, the mechanics of the drafting task have largely been automated and accelerated through the use of such systems. The development of the computer had a major impact on the methods used to design and create technical drawings, making manual drawing almost obsolete, and opening up new possibilities of form using organic shapes and complex geometry.

Today, there are two types of computer-aided design systems used for the production of technical drawings; two dimensions ('2D') and three dimensions ('3D'). 2D CAD systems such as AutoCAD or MicroStation have largely replaced the paper drawing discipline. Lines, circles, arcs and curves are all created within the software. It is down to the technical drawing skill of the user to produce the drawing – though this method does allow for the making of numerous revisions, and modifications of original designs. 3D CAD systems such as Autodesk Inventor or SolidWorks first produce the geometry of the part, and the technical drawing comes from user defined views of the part. This means there is little scope for error once the parameters have been set.

Buildings, Aircraft, ships and cars are now all modelled, assembled and checked in 3D before technical drawings are released for manufacture.

Technical drawing is a skill that is essential for so many industries and endeavours, allowing complex ideas and designs to become reality. It is hoped that the current reader enjoys this book on the subject.

PREFACE

This volume is a book of examples and problems for the study of mechanical drawing. It is intended to be used under the direction of a teacher, although the ambitious and painstaking student can solve many of the problems without such aid. It aims to provide a large selection of typical drawings carefully worked out as examples, each one of which is accompanied by appropriate problems suitable for elementary and advanced work in mechanical drawing. Explanations and directions are, when needed, expressed in simple and direct language, and in the fewest possible words. The principles of the construction of a drawing are shown in the examples in the graphic language of the draftsman.

The book is divided into three parts. The First Part gives in brief outline such matter as the student will need to know before beginning work. It contains, also, a limited number of geometrical constructions which will be found helpful in solving some of the problems given.

The Second Part—the practical work—is divided into four sections, namely: Projections, Developments and Intersections, Isometric Drawings, and Machine Details. Nearly all the example-drawings are fully dimensioned and, where necessary, described by explanatory notes. They are not to be copied but are intended to serve as a guide to the student when making his own drawings for which directions are given in the form of problems. In each section there are many more problems than any one student can work in the amount of time which is usually allotted to the subject of mechanical drawing, and a judicious selection must necessarily be made by the teacher to meet the individual student's requirement. These problems are arranged in the order of their difficulty. The problems given in the beginning of each section are not too difficult for a beginner in mechanical drawing, while those toward the end of the sections will be found suitable for advanced students. Each section begins with problems that are suitable for Junior High Schools, High Schools, and Evening Schools, and advances gradually to problems that are more difficult of solution, suitable for Vocational Schools, Trade Schools, and Colleges. In the first three sections two

v

problems are given with each example-drawing, and in the fourth section, three problems are given.

The problems on Developments and Intersections are suitable as an introductory course for students interested in sheet metal pattern drafting. The problems on Machine Details will form a logical introduction to machine design.

The Third Part of the book contains tables and general information of use to draftsmen, and to which frequent reference should be made when working problems in the second and fourth sections of the Second Part.

The author's not inconsiderable experience as draftsman and designer of machinery, and as teacher of drawing and design, leads him to believe that teachers will find sufficient material in the following pages to enable them to formulate suitable courses for classes in mechanical drawing for all schools where this subject is taught.

The author is indebted to many books on mechanical drawing for helpful suggestions in the preparation of the drawings of this volume, especially to Professor Thomas E. French's "Engineering Drawing." In preparing the manuscript and drawings he gratefully acknowledges much valuable help he has received from Mr. Arthur F. Hopper, director of manual arts, Plainfield, New Jersey; Mr. Frank C. Panuska, instructor in mechanical drawing, and Mr. Ralph Breiling, of Teachers College.

New York,
October, 1917

CONTENTS

PART I.—INTRODUCTION

GENERAL INFORMATION.

LIST OF PLATES

SECTION I

PROJECTIONS

SECTION II

DEVELOPMENTS AND INTERSECTIONS

SECTION III

ISOMETRIC DRAWING

SECTION IV

MACHINE DETAILS

MECHANICAL DRAWING PROBLEMS

PART I

INTRODUCTION

GENERAL INFORMATION

Instruments and Materials.—To obtain good results in mechanical drawing a good set of drawing instruments is necessary. Economy in their purchase is unwise, because drawings made with inferior equipment will be of inferior quality in technique. A good set of drawing instruments with proper care will serve a draftsman's needs almost a lifetime. The following list of instruments and materials comprises a minimum equipment consistent with good work:

A compass with pencil leg, pen leg, and extension bar. Dividers. Bow pencil. Bow pen. Two ruling pens. 45° celluloid triangle. 30° × 60° celluloid triangle. Two celluloid curves. Protractor. 12-inch architect's scale. Drawing board. T-square. Pencils. Sandpaper pencil pointer. Pencil and ink eraser. Two penholders and pens for lettering. Penwiper. Thumb tacks. Cleaning rubber. Drawing paper and drawing ink.

Drawing Lines.—Draw horizontal lines at the upper edge of the T-square blade; never draw lines at its lower edge. Draw vertical lines, and lines making angles 15°, 30°, 45°, 60°, and 75° with the triangles resting against the T-square blade. Lines making angles other than those mentioned are drawn by the aid of the T-square, or the triangle, placed in the desired position.

Fig. 1 shows the method for drawing lines perpendicular to, and also lines making angles of 60°, 30°, and 45° with the T-square blade. The arrows shown on the lines indicate the direction in which the lines should be drawn. Fig. 2 shows methods for drawing lines making angles of 15° and 75° with the T-square blade.

Circles and circular arcs are drawn with the compass, or the bow instruments. Irregular curves are drawn by aid of the celluloid curves, with pencil or ruling pen.

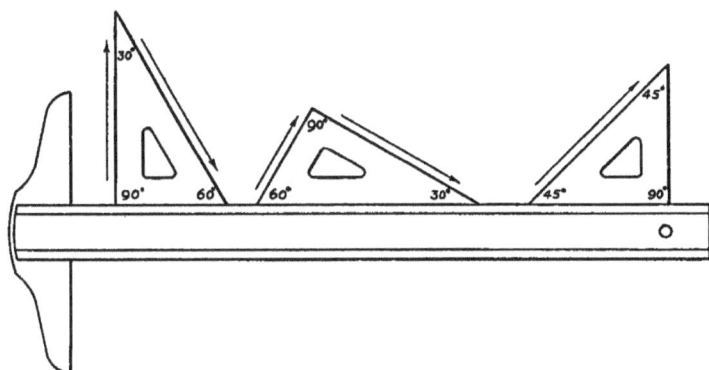

FIG. 1.—Showing vertical and oblique lines.

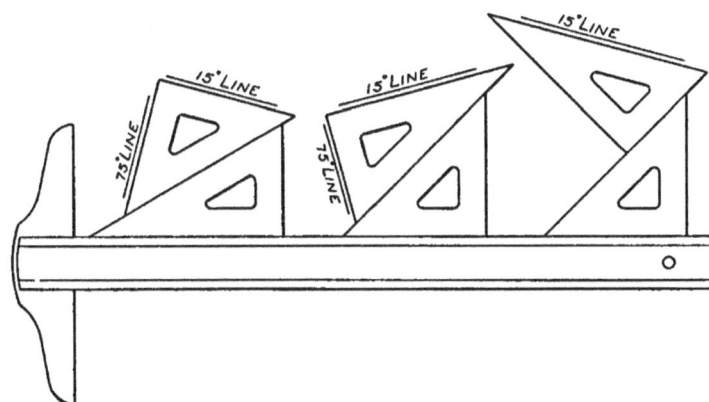

FIG. 2.—Position of triangles for drawing 15° and 75° lines.

Lines Used.—The various lines shown in Fig. 3 are of the kind used by most draftsmen, and may be considered as standard. In drawings which are to be inked or traced, continuous pencil lines may be used where there is no likelihood of an error when inking.

Border Lines.—The object of a border line is to give the drawing a finished appearance. The border line and the trimming line should be drawn in pencil before the drawing itself is begun, and should have the dimensions shown in Fig. 4. When inking, the border line should be the last line drawn.

VISIBLE OUTLINE

HIDDEN OUTLINE

CENTER LINE

PROJECTION LINE

AUXILIARY LINE

EXTENSION LINE

DIMENSION LINE

BORDER LINE

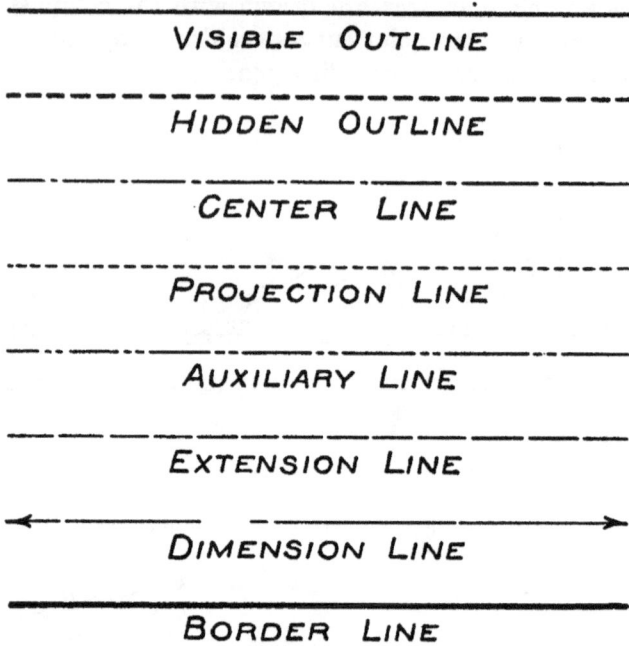

FIG. 3.—Conventional lines.

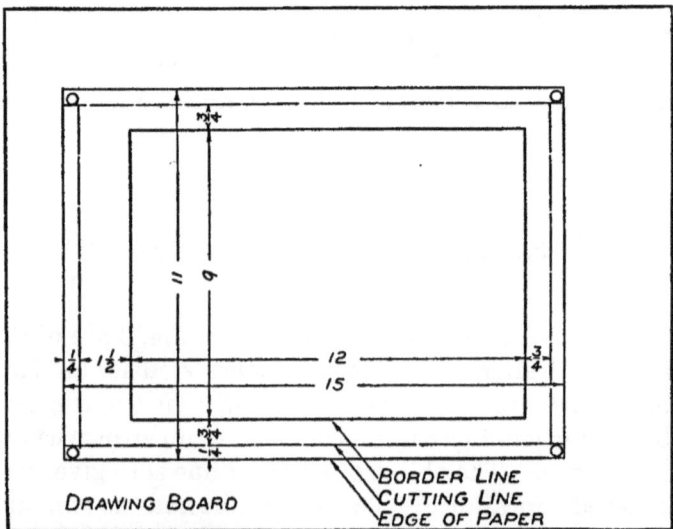

FIG. 4.—Layout of border and cutting lines.

Lettering.—Since the general effect in the appearance of a drawing depends in a large measure on the appearance of its title, notes and dimensions, lettering and figuring form an essential part of the draftsman's work.

The prime requisite for good lettering and figuring is simplicity of style and uniformity of treatment. These are obtained by correctness of form of letters and figures, a uniform inclination and height, and proper spacing. These results must be obtained by accuracy of hand and eye, since no rules can be followed which will be practical for all combinations of letters and words.

Since it is generally difficult for a beginner to letter well, it will be necessary, in order to acquire proficiency and obtain good results, to devote some time to the practice of making letters singly and in combinations to form words.

When placing a title, notes or figures on a drawing the beginner should always remember, no matter how good a drawing may be, or how much time was given to its execution, if the lettering or figuring is hurriedly or carelessly done, the completed drawing will not present a neat appearance.

The "Gothic," or uniform line letters, shown in Fig. 5, find favor with draftsmen and are generally used for mechanical drawings.

Limiting lines should always be drawn to serve as a guide for the height and proper alignment of letters. They are used by the most experienced of letterers. Since it is important that all letters have the same slant, slanting lines should be drawn to serve as a guide to the eye. These lines may be drawn about one-quarter inch apart. See Fig. 6. All limiting and guide lines should be drawn with a wedge-pointed pencil, and very fine so they may be easily erased. Letters and figures are made with a conical-pointed pencil.

Small letters need not necessarily be drawn with the pencil, but may be put in directly with ink. Titles and large letters should always be drawn in pencil before inking.

For titles and large letters a "Hunt's Extra Fine Shot Point Pen," Number 512, and for small letters and figures a "Hunt's Strand Pen," Number 54, will be found suitable. "Leonard's Ball Point Pen," Number 506F, for large letters, and "Gillott's Pen," Number 303, for small letters and figures, also find favor among many draftsmen.

For practical work, the height of letters and figures should be

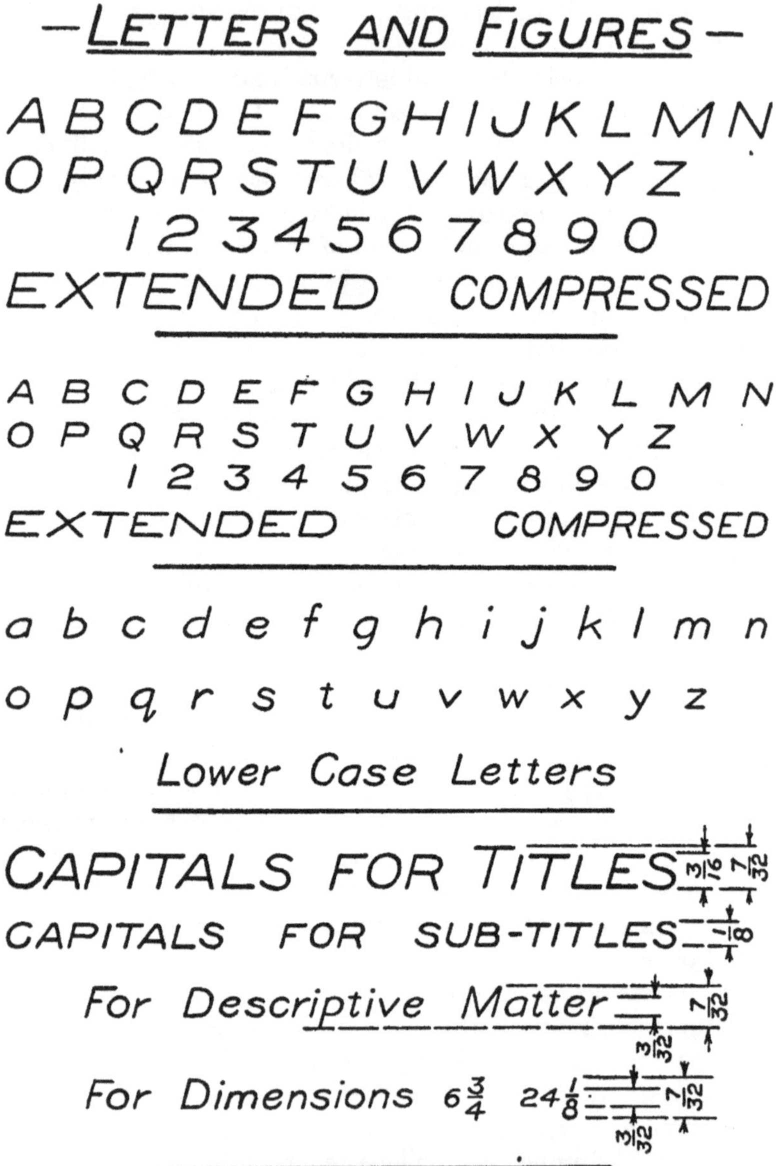

—LETTERS AND FIGURES—

ABCDEFGHIJKLMN
OPQRSTUVWXYZ
1234567890
EXTENDED   COMPRESSED

A B C D E F G H I J K L M N
O P Q R S T U V W X Y Z
1 2 3 4 5 6 7 8 9 0
EXTENDED   COMPRESSED

a b c d e f g h i j k l m n

o p q r s t u v w x y z

Lower Case Letters

CAPITALS FOR TITLES $\frac{3}{16}$ $\frac{7}{32}$

CAPITALS FOR SUB-TITLES $\frac{1}{8}$

For Descriptive Matter $\frac{7}{32}$

For Dimensions 6$\frac{3}{4}$ 24$\frac{1}{8}$ $\frac{3}{32}$ $\frac{7}{32}$ $\frac{3}{32}$

Fig. 5.—Inclined single-stroke letters.

as shown in Fig. 5. The slant may be from 60° to 75°, according to the judgment of the student; or, they may be vertical, if preferable, as shown in the last two lines of Fig. 6.

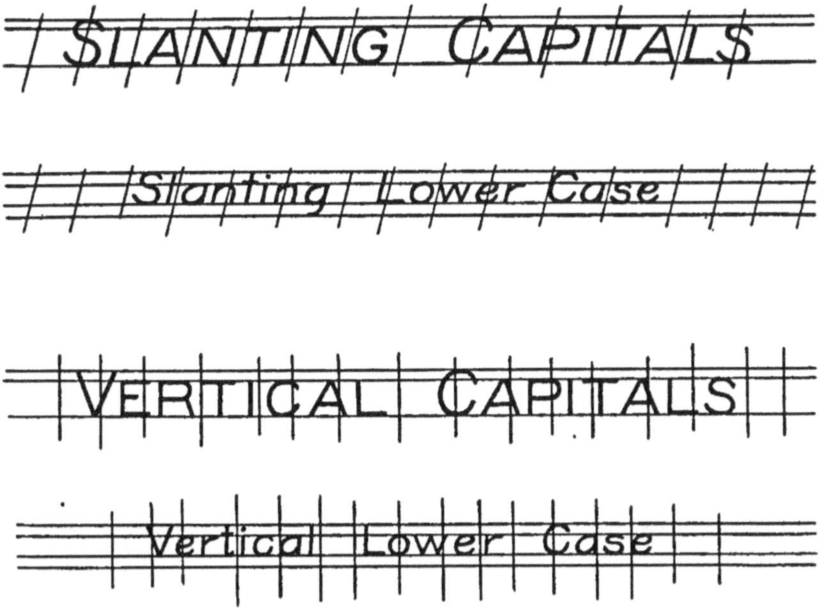

FIG. 6.—Showing use of guide lines.

Dimensions.—To dimension a drawing means to place upon it all measurements required for the construction of the object represented. Fig. 7 shows an object with all dimensions necessary for its construction.

Dimensions are placed in dimension lines which terminate with arrow heads. A break, or space, should be left near but not necessarily in the center of the lines to receive the dimensions. See Fig. 7. Dimensions should not be placed on center lines.

There are several ways for writing dimensions; for instance, four inches may be written as 4 inches, 4 in., 4″, or simply 4, the accent sign (″) being omitted when all dimensions of the object are in inches. The proper use of the sign (″) is shown in Fig. 7. Feet and inches are written thus; 4 ft. 6 in., or, more commonly, 4′–6″; 3′–4½″; 5′–0″, etc. The use or omission of the inch sign, which is omitted in the following drawings, is left to the judgment of the instructor.

Figures for whole numbers should be $\frac{3}{32}$ high, and fractions should be $\frac{7}{32}$ high, over all. The division line of a fraction should be in line with the dimension line. The figures of a fraction should not touch the division line. This requires that each figure of the fraction be a trifle less in height than the whole number. See last line of Fig. 5.

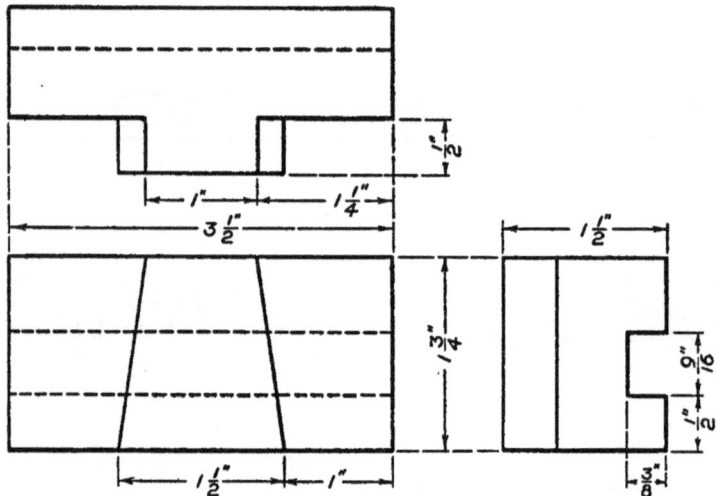

Fig. 7.—Showing a method for dimensioning.

Titles.—The title of a drawing may be placed at the top of the sheet centrally with respect to the vertical lines of the border, and the problem number in the upper left-hand corner, as shown in Fig. 8. The distance marked x should not exceed three-quarters of an inch in any drawing. In some drawings, depending on the problem and the arrangement of views, this distance will necessarily be somewhat less, but in no case should it be less than one-half inch.

In working drawings, titles are always placed in the lower right-hand corner, as shown in Fig. 9, the problem number in the upper left-hand corner being omitted. Fig. 10 shows three forms of general titles. Any one of these forms may be adopted instead of the titles shown in the example drawings.

It is customary in some schools to place titles of practice drawings at the top of the sheet, and titles of working drawings in the lower right-hand corner. A slight readjustment of views in the following drawings, which have their titles on top, will provide space for titles in the lower right-hand corner, if preferable.

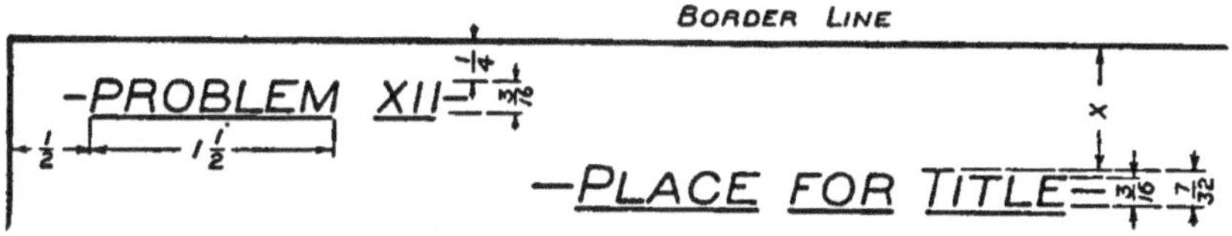

FIG. 8.—Title-form when placed on top.

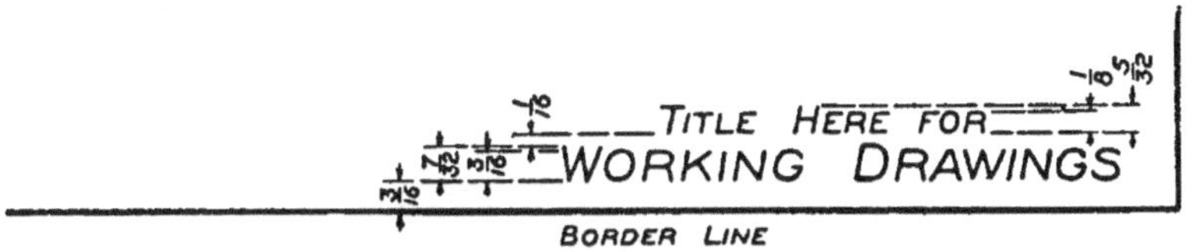

FIG. 9.—Title-form when placed at bottom.

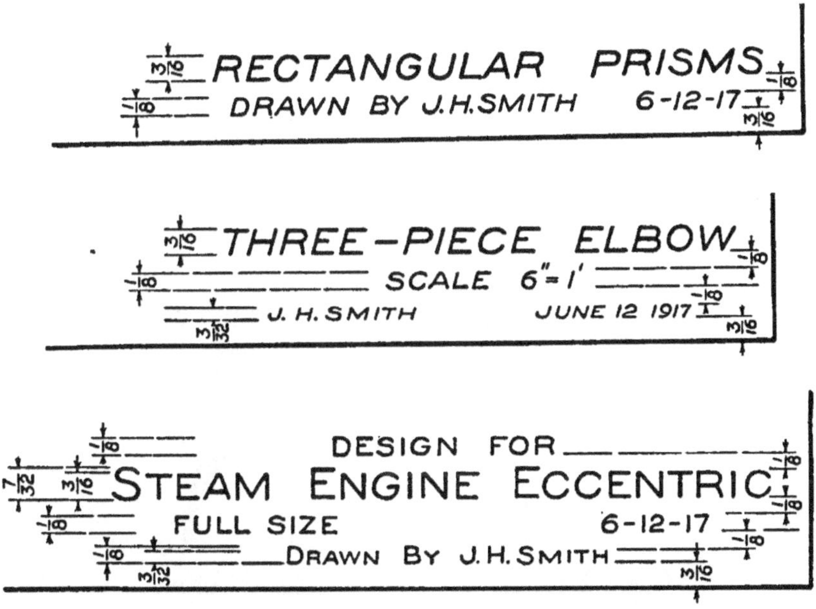

FIG. 10.—Various title-forms.

Before placing a title on a drawing, it would be well to make a trial title on a separate piece of paper, as several attempts may be necessary to produce a satisfactory result. When the trial title is found satisfactory, it can then be copied on the drawing.

Underscore lines drawn to a title, when placed at the top of the drawing, as shown in Fig. 8, are optional with the instructor. Their use, however, adds character and stability to lettering.

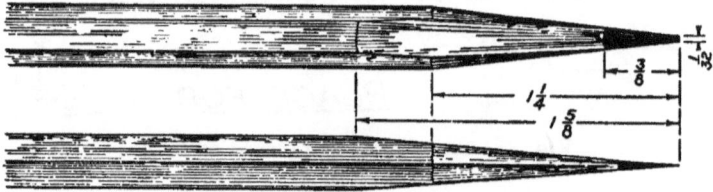

FIG. 11.—Wedge-shaped pencil point.

FIG. 12.—Cone-shaped pencil point.

FIG. 13.—Correct position of pencil.

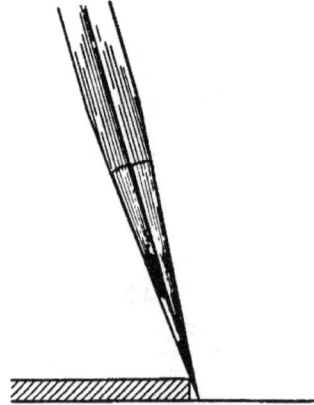

FIG. 14.—Incorrect position of pencil.

Sharpening Pencils.—For wedge-shaped points, remove the wood as shown in Fig. 11, with a sharp knife, or a chisel, exposing the lead which is then sharpened with a sandpaper pencil pointer or a fine file. Fig. 12 shows a cone-pointed pencil.

Fig. 13 shows the correct position of the pencil, relative to the T-square or triangle, for drawing straight and accurate lines. Fig. 14 shows an incorrect position of the pencil.

Penciling.—For good penciling, which is a prerequisite to good inking, two pencils should be used; one 5H for drawing lay-out lines, which are to be drawn fine, and one 4H for drawing lines whose positions and limitations have been determined. These pencils should be sharpened to long slender wedge-shaped points. For pointing off distances, making letters, figures, and arrow heads, a 3H pencil sharpened to a cone-point should be used. Pencils should be sharpened frequently to keep the points in good working condition. All lines should be drawn as fine as is consistent with clearness.

Sequence for Penciling.—A general sequence for penciling, when conditions permit, may be as follows:

1. Draw border and cutting lines.
2. Lay off space required for title.
3. Decide on number of views required.
4. Make a rough, free-hand sketch in note book of views decided upon.
5. Draw horizontal and vertical center lines.
6. Lay out with as few lines as possible the position of each view.
7. Draw limiting horizontal outlines of all views.
8. Draw limiting vertical outlines of all views.
9. Complete all views.
10. Draw dimension lines.
11. Fill in dimensions.
12. Add explanatory notes, if required.

Inking.—When preparing to ink a line do not overload the pen as the ink is likely to flow too freely and thereby cause a blot. Be sure, however, to have enough ink in the pen to finish the line about to be drawn, and always try the pen, after adjustment for width of line, on a piece of paper. When finished do not lay the pen aside without cleaning.

Fig. 15 shows the correct position of the ruling pen for inking, relative to the T-square, triangle, or irregular curve. The ruling pen should not be used for drawing free-hand lines.

Sequence for Inking.—A sequence for inking is more necessary than one for penciling, due to the necessity of changing instruments and waiting for ink to dry. A good order for inking is as follows:

1. Small circles and circular arcs with the bow pen.
2. Large circles and circular arcs with the compass.
3. Irregular curves.
4. Horizontal lines beginning with the uppermost.
5. Vertical lines beginning with those at the left-hand end.
6. All 30°, 60°, and 45° lines.
7. Other oblique lines.
8. Horizontal and vertical center lines.
9. Extension and dimension lines.
10. Put in dimensions, arrow heads, and explanatory notes.
11. Section lines.
12. Border line.

In some drawings it may be desirable to ink center lines first.

Use of Lines.—The diagram drawing shown in Fig. 16 illustrates the correct use of the lines shown in Fig. 3. It also shows the correct and incorrect methods of forming junctions of full and hidden lines and of making arrow heads. The lengths of dashes and spaces of broken lines, and the size of arrow heads, are to be estimated by eye and should as closely resemble those shown in Fig. 3 as conditions permit.

The drawing of projection lines is recommended for all practice drawings. In working drawings projection lines should be omitted.

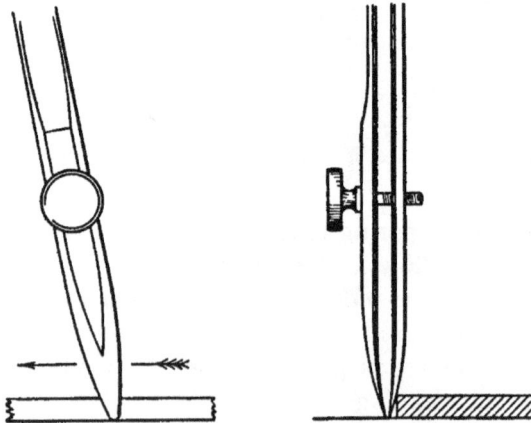

Fig. 15.—Correct position of ruling pen.

Location of Views.—The amount of space to be occupied by the title, and the size and location of each view of an object about to be drawn, should be determined before the drawing is begun. This information can be obtained beforehand by making a free-

hand sketch, preferably in a note book kept for the purpose, and
to a scale if necessary, of the desired views, the amount of space
between the views, and the distance of each view from the horizontal and the vertical lines comprising the border line.

In many cases, if the drawing is quite simple, a few lines drawn
very lightly will serve as a preliminary lay-out, and will permit
of a readjustment of views should the first trial be unsuccessful.

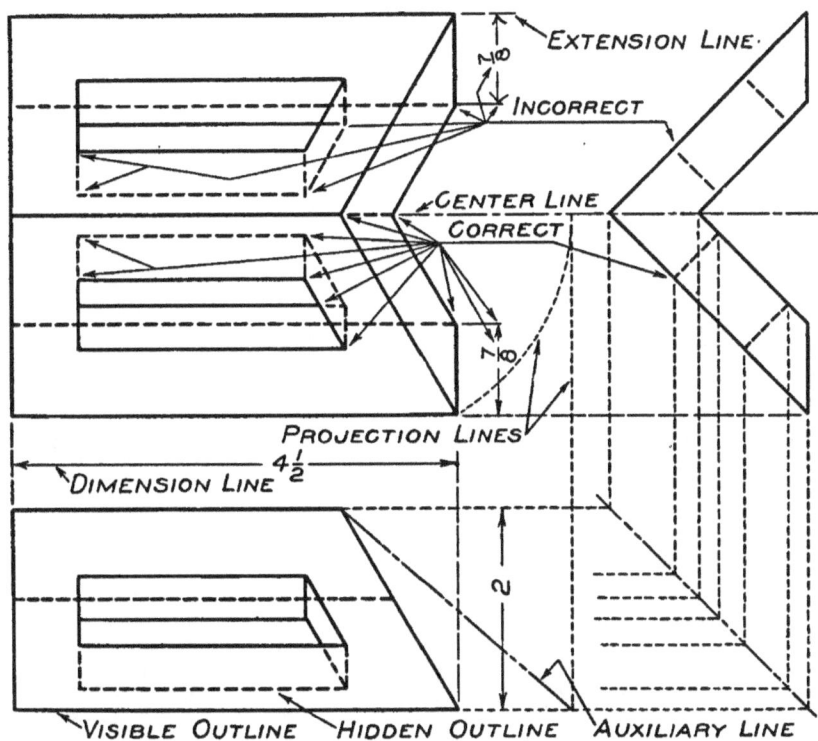

Fig. 16.—Correct and incorrect junctures of lines.

To obtain a pleasing appearance, the various views should be
arranged and placed in such a way as to utilize the available
space on the sheet to the best advantage.

Working Drawings.—A working drawing, also called a shop
drawing, is one which will impart such definite and unmistakable
information in the form of a graphic representation with all necessary dimensions, notes and explanatory matter, the kinds of
materials to be used and methods of finishing, as is required by a
workman for the construction of the object represented.

GEOMETRIC CONSTRUCTIONS

To Bisect a Given Angle (Fig. 17).

Let ABC be the given angle. With B as center and any suitable radius describe arc ab. With a and b as centers and any suitable radius describe arcs intersecting at c. Draw Bc, the bisector of the angle.

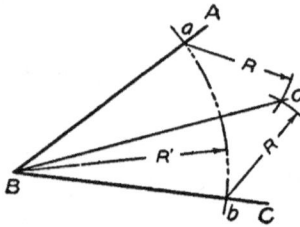
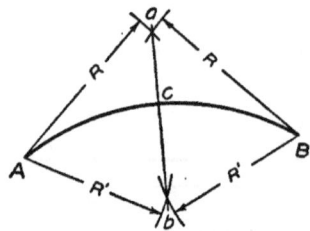

FIG. 17.   FIG. 18.

To Bisect a Given Arc (Fig. 18).

Let AB be the given arc. With A and B as centers and any suitable radius describe arcs intersecting at a. With A and B as centers, and the same or other suitable radius, describe arcs intersecting at b. Draw ab intersecting the arc at c, the required point.

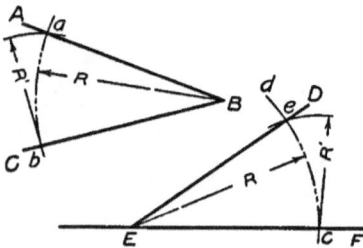
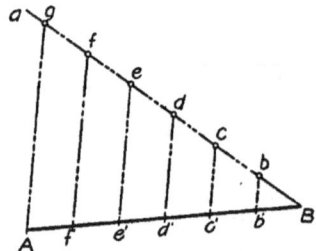

FIG. 19.   FIG. 20.

To Set off an Angle Equal to a Given Angle from a Point on a Given Line (Fig. 19).

Let ABC be the given angle and E the given point on line EF. With B as center and any suitable radius draw arc ab. From E with the same radius draw arc cd. With radius ab and c as center, intersect cd in e. Draw Ee. Then angle DEF equals angle ABC.

To Divide a Given Line into Any Number of Equal Parts (Fig. 20).

Let AB be the given line, and six the required number of parts. Draw Ba, at any angle to AB. With any convenient length lay off on Ba the required number of parts, giving points b, c, d, e, f, g. From the points on Ba, and parallel to gA, draw lines intersecting AB at b', c', d', e', f', giving the required parts.

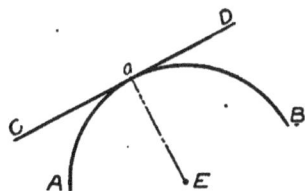

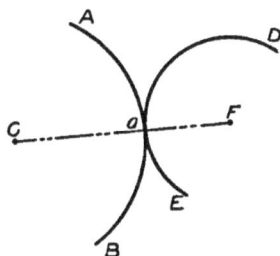

FIG. 21.   FIG. 22.

To Find the Point of Tangency of a Given Circular Arc and a Given Straight Line (Fig. 21).

Let AB be the given arc of center E; and CD, the given line. From E draw a perpendicular to CD. The intersection a will be the point of tangency.

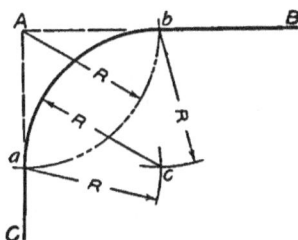

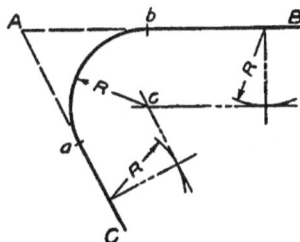

FIG. 23.   FIG. 24.

To Find the Point of Tangency of Two Given Circular Arcs (Fig. 22).

Let AB of center C and ED of center F be the given arcs. Draw CF. The intersection a will be the point of tangency.

To Draw an Arc of Given Radius Tangent to Two Straight Lines Meeting at Right Angles (Fig. 23).

Let AB and AC be the given lines and R the given radius. With A as center and R as radius draw an arc intersecting the given lines in a and b. With a and b as centers and the same radius draw arcs intersecting at c. With c as center and the same radius draw the required arc ab. Points a and b are the points of tangency of the arc and the given lines.

To Draw an Arc of Given Radius Tangent to Two Intersecting Straight Lines (Fig. 24).

Let AB and AC be the given lines, and R the given radius. At a distance R draw parallels to AB and AC, intersecting at c. With c as center and radius R draw the required arc. Points a and b are the points of tangency.

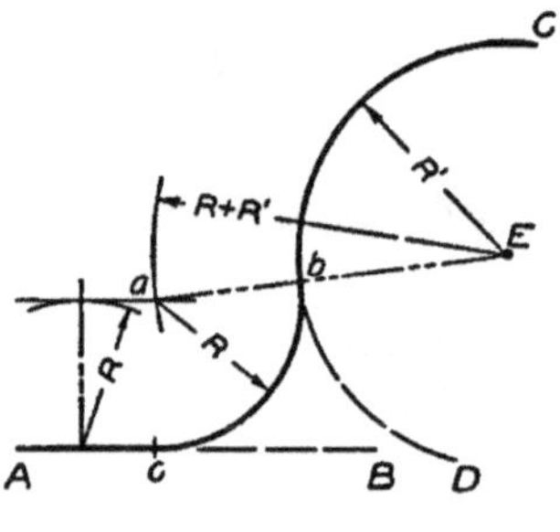

FIG. 25. 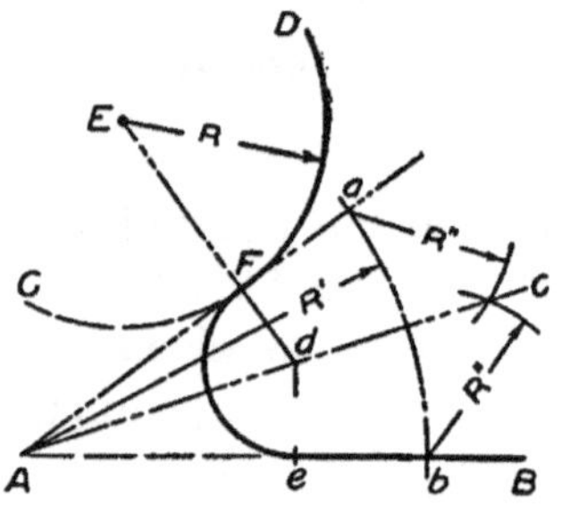 FIG. 26.

To Draw an Arc of Given Radius Tangent to a Given Straight Line and a Given Circular Arc (Fig. 25).

Let AB be the given line, CD the given circular arc of radius R', and R the given radius. With E as center and radius $R + R'$ draw an arc; also draw a line parallel to AB at distance R, intersecting the arc at a. With a as center and radius R draw the required arc. Points b and c are the points of tangency.

NOTE.—The point of tangency b lies on a straight line joining centers a and E, and the point of tangency c lies at the foot of a perpendicular drawn from a to AB.

To Draw a Circular Arc Tangent to a Given Straight Line and Tangent at a Point on a Given Circular Arc (Fig. 26).

Let AB be the given line and F the point on the given arc CD of radius R.   Draw a line tangent at F intersecting line AB at A. With A as center and radius R' draw arc ab.   With a and b as centers and radius R'' draw arcs intersecting at c.   Draw Ac; also draw a straight line from E through F giving point d. With d as center and radius dF draw the required arc.   Points F and e are the points of tangency.

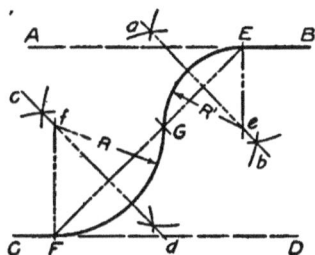

Fig. 27.

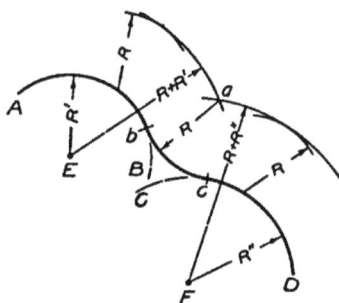

Fig. 28.

To Connect Two Given Parallel Lines with a Compound Curve and Tangent at Given Points (Fig. 27).

Let AB and CD be the given lines; E and F the given points. Connect E and F by a straight line.   Assume any point, as G. Draw ab and cd, the bisectors of EG and GF, respectively.   At E and F erect perpendiculars giving points e and f.   With center f and radius R, equal to fF, draw arc GF.   With center e and radius R', equal to eE, draw arc GE, completing the required curve.   Points E and F are the points of tangency.   Point G is the point of tangency of the two arcs.

To Draw a Circular Arc of Given Radius Tangent to Two Given Circular Arcs (Fig. 28).

Let R be the radius of the given arc; and R' and R'' the radii of the given circular arcs.   With E and F as centers, and radii $R + R'$ and $R + R''$, draw arcs intersecting at a.   With a as center and radius R draw the required arc.   Points b and c, on straight lines drawn from a to E and a to F, respectively, are the points of tangency.

To Construct a Regular Polygon of Any Number of Sides Within a Circle of Given Diameter (Fig. 29).

Let *ABCD* be the given circle. Divide *AC* into as many equal

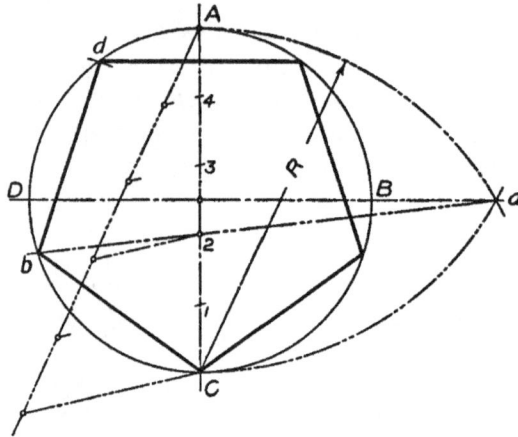

FIG. 29.

parts as the polygon is to have sides, in this case five. With *A* and *C* as centers, describe arcs of radius *AC*, intersecting at *a*. A line drawn through *a*2, intersecting the circle at *b*, determines

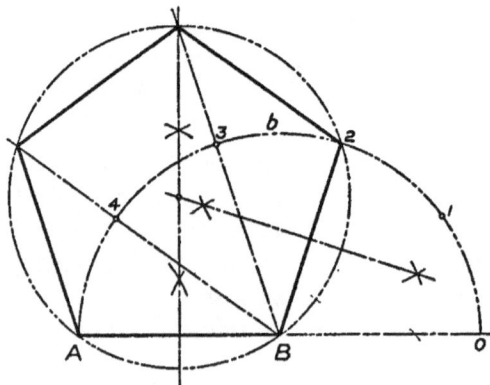

FIG. 30.

the length *bC*, one side of the polygon. With *bC* as radius and *b* as center describe a small arc giving point *d*. The other points are found similarly.

2

To Construct a Regular Polygon of Any Number of Sides on a Line of Given Length (Fig. 30).

Let AB be the given length. With B as center and AB as radius draw the semi-circle Abo. Divide the semi-circle into as many equal parts as the polygon is to have sides, in this case five. Draw $B2$. Bisect AB and $B2$ to find the center of the circumscribing circle. Draw lines from B through 3 and 4. The intersection of these lines with the circumscribed circle determines the vertices of the polygon.

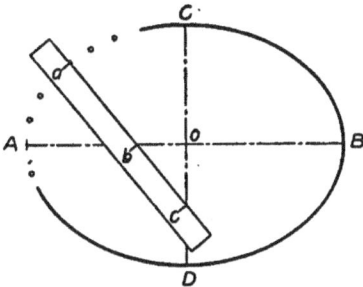

Fig. 31.

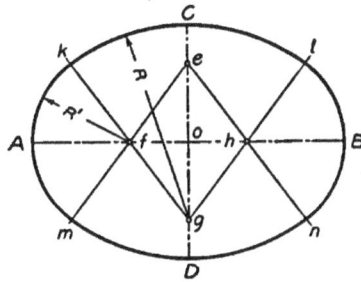

Fig. 32.

To Draw an Ellipse, the Major and Minor Axes Being Given (Fig. 31).

Let AB be the major axis and CD the minor axis. Lay off on the edge of a straight piece of paper the distance ac equal to Ao, the semi-major axis. Also lay off the distance ab equal to Co, the semi-minor axis. Place the paper so that b coincides with the major axis, and c coincides with the minor axis, or the minor axis produced; then a will give one point on the required ellipse. Locate as many points as are necessary to draw a smooth curve.

To Draw an Approximate Ellipse, the Major and Minor Axes Being Given (Fig. 32).

Let AB be the major axis and CD the minor axis. On the minor axis lay off oe and og equal to the difference between the major and the minor axes. On the major axis lay off of and oh equal to three-fourths of oe Draw ef, eh, gf, and gh, all produced. With center e draw arc mDn. With center g draw arc kCl. With center f draw arc kAm. With center h draw arc lBn, completing the required ellipse.

CONVENTIONAL SCREW THREADS

Screw Threads.—Fig. 33 shows a method for drawing conventional single thread screws. Draw two lines indicating the diameter. On one line lay off spaces equal to the pitch. Bisect one space and draw line *ab*. From *b* draw an inclined line of half

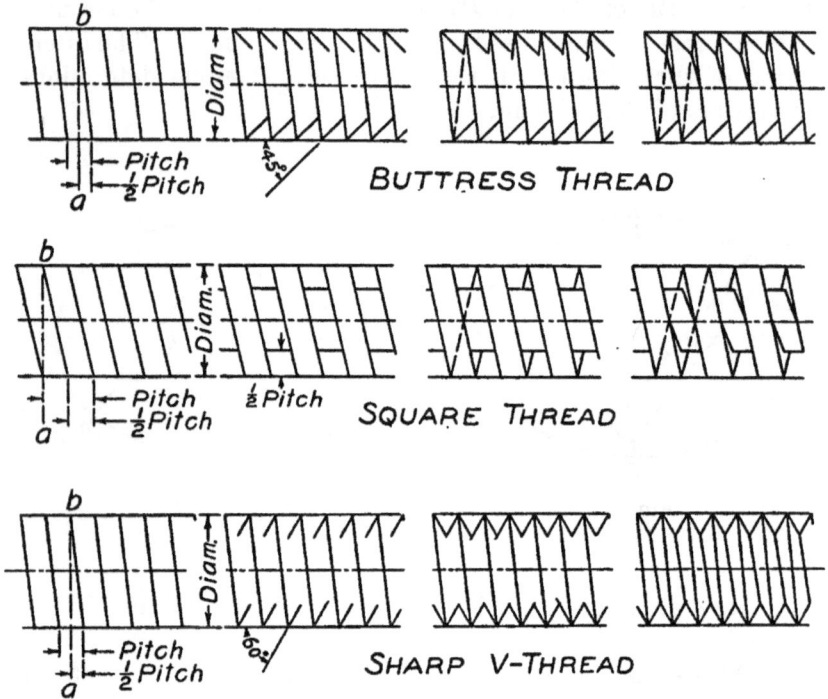

BUTTRESS THREAD

SQUARE THREAD

SHARP V-THREAD

Fig. 33.—Showing steps for drawing conventional threads.

the pitch. From the spaces laid off on one line draw parallels to the inclined line. Complete the thread, following the steps suggested in the figure. These conventions are used for large screw threads.

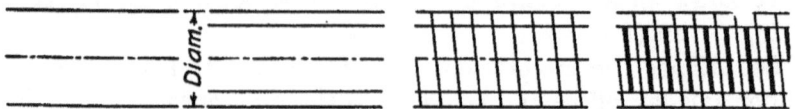

Fig. 34.—Common conventional thread.

For screw threads under three-quarter inch diameter the convention shown in Fig. 34 may be used. The spacing for pitch should be estimated by the eye.

INTERSECTION OF TWO CYLINDERS

To Find the Line of Intersection of Two Cylinders with Axes in the Same Plane and at Right Angles to Each Other (Fig. 35).

Let A, B, and C be the front, end, and top views, respectively. Points on the line of intersection may be found by the intersection of elements in the surface of one cylinder, with elements in the

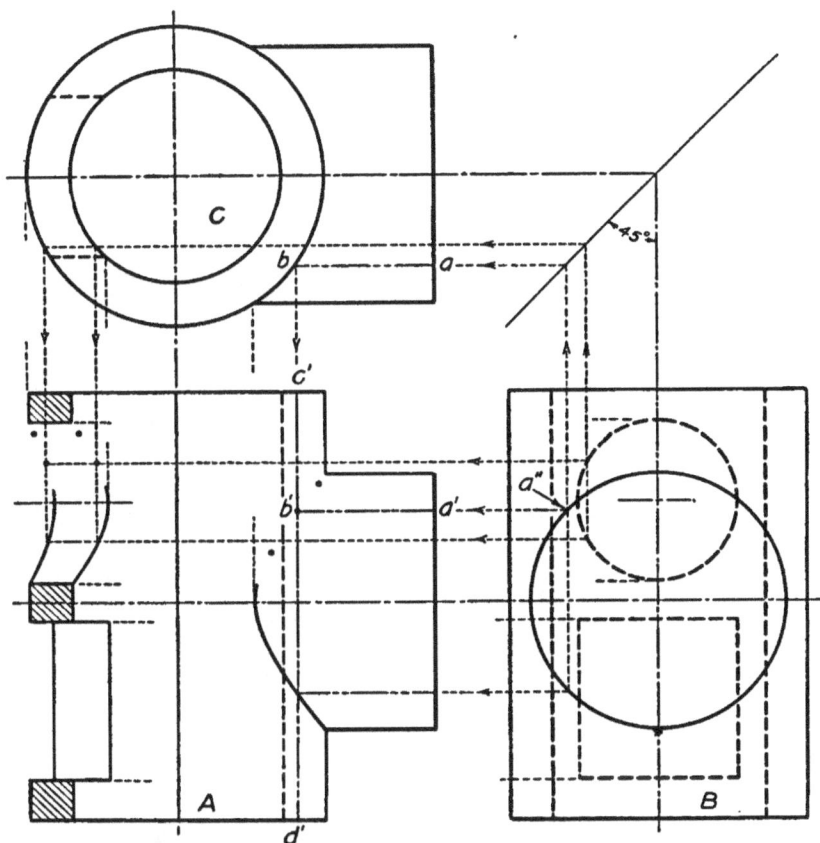

FIG. 35.—Intersection by elements method.

surface of the other. Let a'', an assumed point, be the end view of an element, in the surface of the horizontal cylinder, which projected to the front and the top views gives $a'b'$ and ab, the front and the top views of the element. Let b be the top view of an element, in the surface of the vertical cylinder, which projected to the front view gives $c'd'$, the front view of the element. The point of intersection of these elements gives b', one point on the required line of intersection. Additional points are found similarly.

The method of finding the lines of intersection of the hollow vertical cylinder and the two holes cut through its thickness is evident from the figure.

PLANE INTERSECTION OF A SOLID

To Find the Lines of Intersection of a Surface of Revolution Cut by Two Planes at Right Angles to Each Other and Parallel to the Axis (Fig. 36).

Let ab and $a'b'$ be the projections of the axis, pp and $p'p'$ the end view of the cutting planes, and cd the circular-arc outline of the surface. A transverse section at e, an assumed point on the

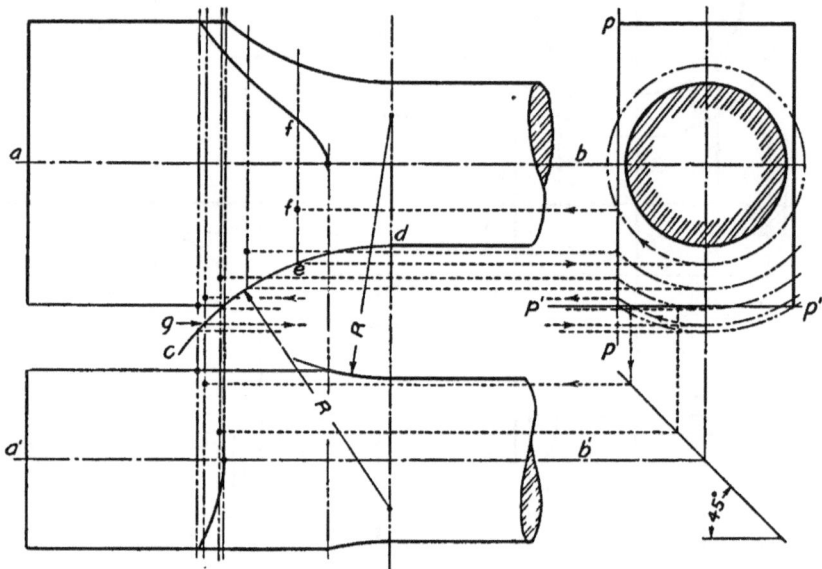

FIG. 36.—Intersection by cutting-plane method.

curve, shown in the end view as a circle, is cut by plane pp. The intersecting points of the circle and plane projected back to the top view give ff, points on the required line of intersection. A transverse section at g, another assumed point, shown in the end view by the arc of a circle, is cut by planes pp and $p'p'$. The intersecting points of the arc and planes projected back to the top and the front views give additional points on the line of intersection. Other points are found in a similar manner. Through these points, smooth curves are drawn.

PART II
EXAMPLES AND PROBLEMS

DEFINITIONS

PLANES OF PROJECTION

For example see Plate 5

The **Ground Line,** designated in the drawing as *GL*, shows the division of two **Planes.** The surface above this line is called the **Horizontal Plane,** or *H*; the surface below the line is called the **Vertical Plane,** or *V*.

The line marked *G'L'* is called an **Auxiliary Ground Line** and shows the line of intersection of the vertical plane and of an **Auxiliary Horizontal Plane.**

PROJECTIONS ON THREE PLANES

For example see Plate 25

When drawing three views of an object, a **Side,** or **Profile Plane** is used in addition to the horizontal and vertical planes. The line *abd* is the ground line, while the line *cbe* is the **Profile Plane Trace.** The surface bounded by *abc* is the horizontal plane. The surface bounded by *abe* is the vertical plane and the surface bounded by *ebd* is the side, or profile plane, frequently designated as *P*.

PROJECTIONS ON AUXILIARY PLANES

For example see Plate 29

Projections on **Auxiliary Planes** are frequently made to show the true shape of oblique surfaces; that is, of surfaces which are not parallel to any one of the regular planes of projection. Auxiliary planes are generally perpendicular to *H* and inclined to *V*, or perpendicular to *V* and inclined to *H*.

The surface bounded by *ebd* is the auxiliary plane, while the line *be* is the **Auxiliary Plane Trace.**

23

SECTION I

PROJECTIONS

PRISMS

This drawing shows the top views, also called plans, or horizontal projections, and the front views, also called front elevations, or vertical projections, of four right prisms. The front views, since the prisms are all the same height and same width, are alike. The top views, which show their outline or shape, cannot be determined from the front views; therefore, two views are necessary to represent such objects completely.

Problem 1.—Make a drawing of four prisms similar to those shown. Let A = 3 inches, B = 2 inches, and C = 1 inch.

Problem 2.—Draw top and front views of four prisms similar to those shown. Let A = $3\frac{1}{4}$ inches, B = $1\frac{7}{8}$ inches, and C = $1\frac{1}{4}$ inches.

TAPERED OBJECTS

This drawing shows the top and the front views of four tapered objects. The front views are alike while the top views differ. The first object is a wedge; the fourth, a cone.

It will be observed that the front view of a wedge, and the front view of a cone may be exactly alike, although the objects are radically different, as shown by their top views.

Problem 1.—Draw top and front views of four objects similar to those shown. Let A = $2\frac{3}{8}$ inches, B = $1\frac{1}{2}$ inches, and C = $\frac{3}{4}$ inch.

Problem 2.—Make a drawing showing top and front views of four objects similar to those shown. Let B = $1\frac{3}{8}$ inches, C = 1 inch, and A = 2, $2\frac{1}{8}$, $2\frac{1}{2}$, and $2\frac{3}{4}$ inches, respectively.

PLATE 1

PRISMS

Top Views or Plans or Horizontal Projections

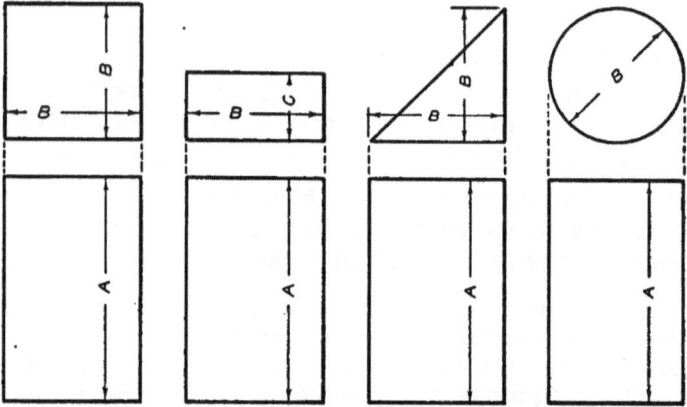

Front Views or Front Elevations or Vertical Projections

TAPERED OBJECTS

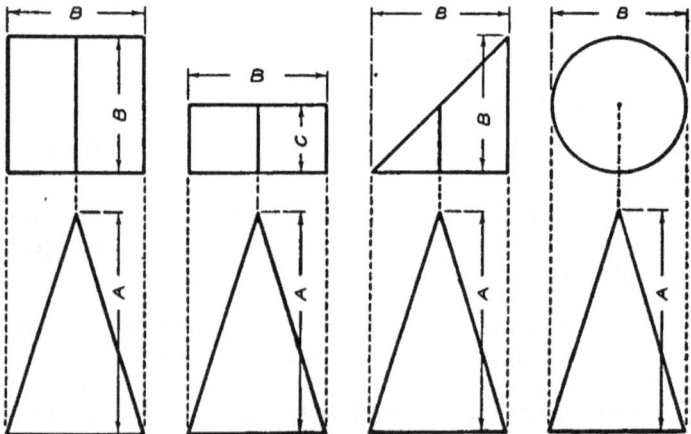

PLATE 2

CIRCULAR OBJECTS

This drawing shows four circular objects whose top views are alike and whose front views vary considerably.

Problem 1.—Draw top and front views of objects similar to those shown and of the following dimensions:

	A	B	C	D	E
First object........:...	$1\frac{3}{4}$	$\frac{7}{8}$	3		
Second object.......	$1\frac{3}{4}$	$\frac{7}{8}$	$3\frac{1}{4}$	$\frac{3}{4}$	
Third object........	$1\frac{3}{4}$	$\frac{7}{8}$	$3\frac{1}{4}$	$1\frac{1}{4}$	1
Fourth object.......	$1\frac{3}{4}$	$\frac{7}{8}$	$3\frac{3}{8}$	$1\frac{1}{2}$	$\frac{3}{4}$

Problem 2.—Draw top and front views of objects similar to those shown and of the following dimensions:

	A	B	C	D	E
First object..	$1\frac{3}{4}$	$\frac{7}{8}$	$3\frac{3}{16}$		
Second object.......	$1\frac{13}{16}$	$\frac{15}{16}$	$3\frac{5}{16}$	$\frac{7}{8}$	
Third object........	$1\frac{7}{8}$	1	$3\frac{7}{16}$	$1\frac{5}{32}$	$1\frac{1}{4}$
Fourth object.......	$1\frac{15}{16}$	$1\frac{1}{16}$	$3\frac{9}{16}$	$1\frac{5}{8}$	$\frac{7}{8}$

GEOMETRIC OBJECTS

This drawing shows four objects whose top views are similar but whose front views differ.

Problem 1.—Make a drawing showing top and front views of objects similar to those shown and of the following dimensions:

	A	B	C	D	E	F	G	H
First object.....	$1\frac{1}{2}$	$3\frac{3}{4}$	$\frac{1}{2}$	$1\frac{1}{8}$	$2\frac{1}{2}$			
Second object....	$1\frac{5}{8}$	$3\frac{1}{4}$	$\frac{5}{8}$	$1\frac{1}{4}$	$\frac{1}{2}$	$2\frac{1}{4}$		
Third object....	$1\frac{1}{4}$	$3\frac{3}{8}$	$\frac{3}{4}$	$1\frac{3}{8}$	$\frac{3}{4}$			
Fourth object....	$1\frac{7}{8}$	3	$\frac{7}{8}$	$1\frac{1}{2}$	1	$\frac{7}{8}$	$1\frac{3}{8}$	$1\frac{5}{8}$

Problem 2.—Make a drawing showing top and front views of objects similar to those shown and of the following dimensions:

	A	B	C	D	E	F	G	H
First object.....,	$1\frac{7}{8}$	$3\frac{1}{4}$	1	$1\frac{5}{8}$	$2\frac{1}{8}$			
Second object....	$1\frac{3}{4}$	$3\frac{1}{4}$	$\frac{7}{8}$	$1\frac{1}{2}$	$\frac{5}{8}$	2		
Third object....	$1\frac{5}{8}$	$3\frac{3}{8}$	$\frac{3}{4}$	$1\frac{3}{8}$	$1\frac{11}{16}$			
Fourth object....	$1\frac{1}{2}$	$3\frac{1}{2}$	$\frac{3}{4}$	$1\frac{1}{4}$	$\frac{1}{2}$	$\frac{3}{4}$	$2\frac{1}{2}$	$2\frac{11}{16}$

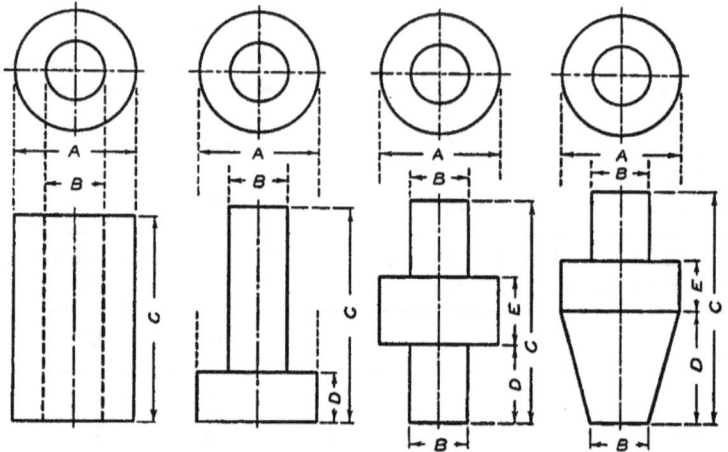

CIRCULAR OBJECTS

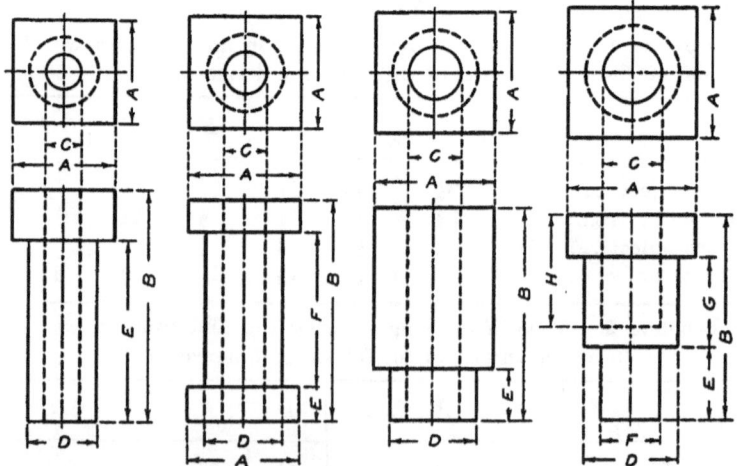

GEOMETRIC OBJECTS

PLATE 4

GEOMETRIC OBJECT

This drawing shows the top and the front views of an object in three positions in relation to *GL*.

The top views show the object in contact with *GL*, therefore in contact with *V*. See page 23.

Problem 1.—Draw top and front views of the object in positions similar to those shown and of the following dimensions:

	A	B	C	D	E	θ	φ
First position.........	2	$1\frac{1}{2}$	$2\frac{3}{4}$	$\frac{3}{8}$	$\frac{5}{16}$		
Second position........	2	$1\frac{1}{2}$	3	$\frac{3}{8}$	$\frac{5}{16}$	30°	
Third position.........	2	$1\frac{1}{2}$	$3\frac{1}{4}$	$\frac{3}{8}$	$\frac{5}{16}$		45°

Problem 2.—Make a drawing showing top and front views of the object in positions similar to those shown and of the following dimensions:

	A	B	C	D	E	θ	φ
First position.........	$1\frac{7}{8}$	$1\frac{3}{8}$	$2\frac{1}{4}$	$\frac{5}{16}$	$\frac{1}{4}$		
Second position........	2	$1\frac{1}{2}$	3	$\frac{3}{8}$	$\frac{5}{16}$	15°	
Third position.........	$2\frac{1}{8}$	$1\frac{5}{8}$	$3\frac{1}{4}$	$\frac{7}{16}$	$\frac{3}{8}$		45°

GEOMETRIC SOLID

This drawing shows the top and the front views of a triangular solid in three positions.

Problem 1.—Draw top and front views of a similar solid having the following dimensions and positions:

	A	B	C	D	E	θ	φ
First position.........	$2\frac{1}{4}$	$1\frac{3}{4}$	$2\frac{7}{8}$	$1\frac{1}{4}$	$\frac{7}{8}$		
Second position........	$2\frac{3}{8}$	$1\frac{3}{4}$	3	$1\frac{1}{4}$	1	45°	
Third position.........	$2\frac{1}{2}$	$1\frac{3}{4}$	$3\frac{3}{8}$	$1\frac{1}{4}$	$1\frac{1}{8}$		30°

Problem 2.—Make a drawing showing top and front views of a similar solid having the following dimensions and positions:

	A	B	C	D	E	θ	φ
First position.........	$2\frac{1}{4}$	2	$3\frac{1}{4}$	$1\frac{5}{8}$	1		
Second position..	$2\frac{3}{8}$	$1\frac{7}{8}$	3	$1\frac{1}{2}$	$\frac{3}{4}$	15°	
Third position.........	$2\frac{1}{2}$	$1\frac{3}{4}$	$2\frac{3}{4}$	$1\frac{3}{8}$	$\frac{1}{2}$		75°

PLATE 5

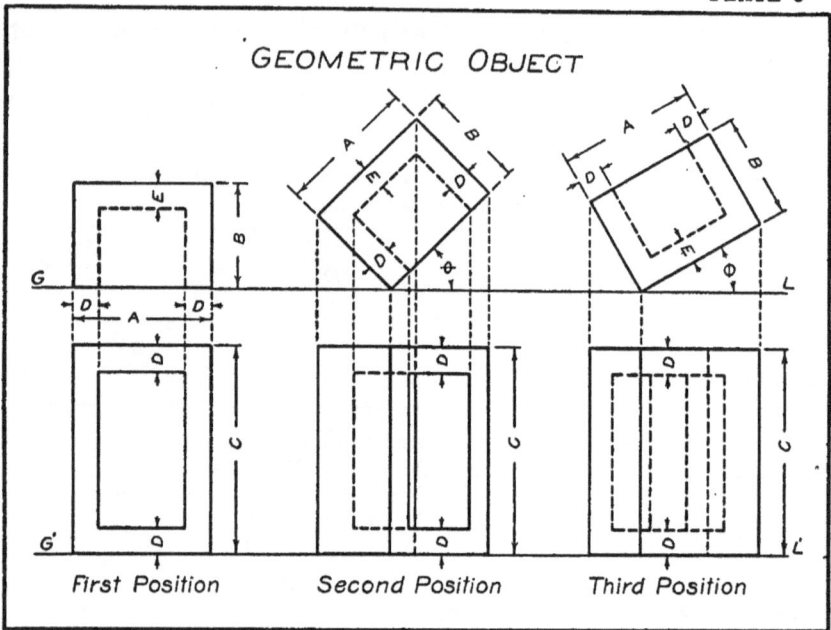

GEOMETRIC OBJECT

First Position   Second Position   Third Position

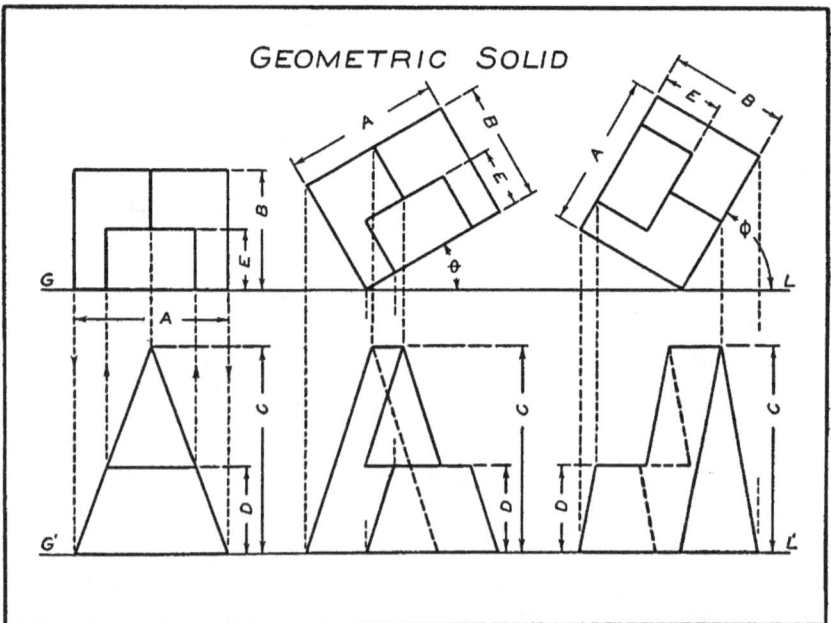

GEOMETRIC SOLID

PLATE 6

PROJECTION OF LETTER (V)

This drawing shows the top and the front views of a letter consisting of horizontal, vertical, and oblique lines, in three positions relative to V.

In solving the following problems draw the front view of the letter in its first position and project the top view.   Transpose the top view to the required positions and project the front views.

Problem 1.—For the first position, draw the front view of the letter as shown; project the top view when removed one inch from V.   For the second position, let one corner be in contact with V, and $\theta = 30°$.   For the third position show the nearest corner removed $\frac{1}{4}$ inch from V, and $\phi = 15°$.

Problem 2.—Draw front and top views, as explained in Problem 1, when the letter is inverted.   Insert a cross-bar and change the letter to **A**. Show hidden lines in the front view of the third position.

PROJECTION OF LETTER (K)

This drawing shows the top and the front views of a letter consisting of horizontal, vertical, and oblique lines in three positions.

For drawing the following problems read the instructions given for letter **V**.

Problem 1.—Draw front and top views in positions similar to those shown and of the following dimensions:

A	B	C	D	E	F	G	θ	ϕ
$3\frac{5}{8}$	$3\frac{1}{4}$	3	$\frac{3}{4}$	$\frac{3}{4}$	$\frac{1}{4}$	$2\frac{1}{8}$	30°	45°

Show all hidden lines in the front view of the second position.

Problem 2.—Using the letter **K** as a guide, design the letter **X** and draw views in positions similar to those shown for **K** and of general dimensions, as follows:

A	B	C	D	E	F	G	θ	ϕ
$3\frac{3}{4}$	$3\frac{1}{2}$	$3\frac{1}{4}$	1	$\frac{3}{4}$	$\frac{1}{4}$	$2\frac{1}{4}$	15°	30°

Show all hidden lines in the front view of the third position.

PLATE 7

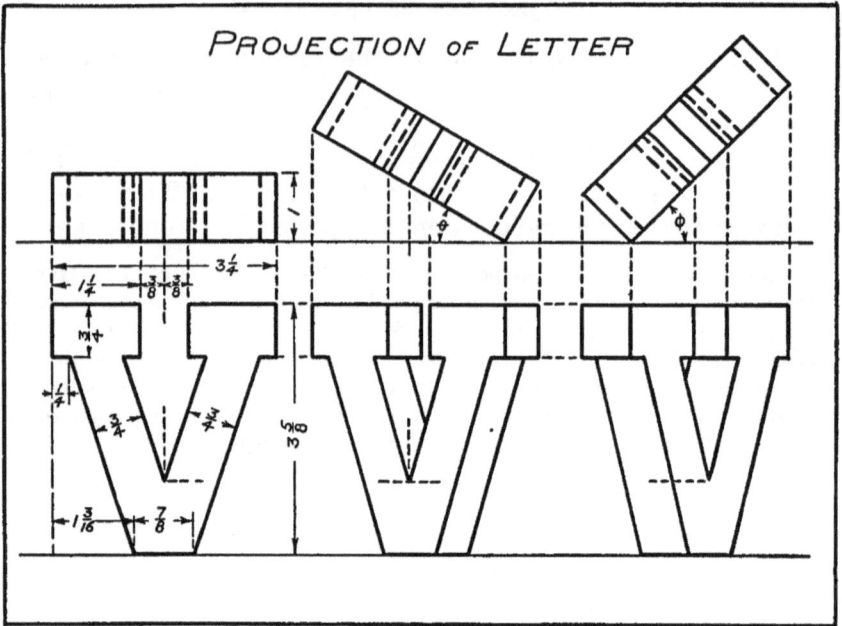

PROJECTION OF LETTER

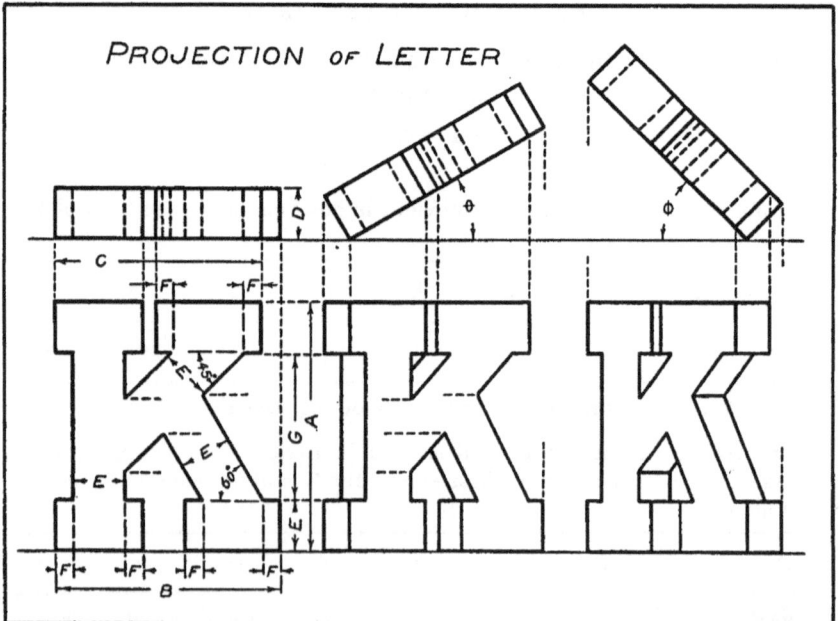

PROJECTION OF LETTER

PLATE 8

HOLLOW CYLINDER

This drawing shows the top and the front views of a hollow cylinder in three positions.

For the first position, draw front and top views as shown. Divide the front view, giving points 1', 2', 3', etc. Project these points to the top view. For the second and the third positions, transpose the top view and locate points for the front views by projecting lines as shown.

Problem 1.—Draw the cylinder in positions similar to those shown and having the following dimensions:

A	B	C	θ	ϕ
3	$1\frac{3}{4}$	$1\frac{1}{8}$	30°	60°

Complete the second view and show hidden lines.

Problem 2.—Draw the cylinder in positions similar to those shown and having the following dimensions:

A	B	C	θ	ϕ
$2\frac{3}{4}$	2	1	15°	30°

Complete the second view and show hidden lines in the third position.

GEOMETRIC SOLID

This drawing shows the top and the front views of a solid in three positions.

Problem 1.—Draw top and front views of the solid in positions similar to those shown and having the following dimensions:

A	B	C	D	E	θ	ϕ
2	$2\frac{1}{4}$	$3\frac{1}{8}$	$1\frac{3}{4}$	1	30°	45°

Complete the second view and show hidden lines.

Problem 2.—Draw views similar to those shown when the solid has the following dimensions:

A	B	C	D	E	θ	ϕ
$1\frac{7}{8}$	$2\frac{1}{8}$	$2\frac{15}{16}$	$1\frac{1}{2}$	$\frac{3}{4}$	15°	75°

Complete the second view and show hidden lines in the third position.

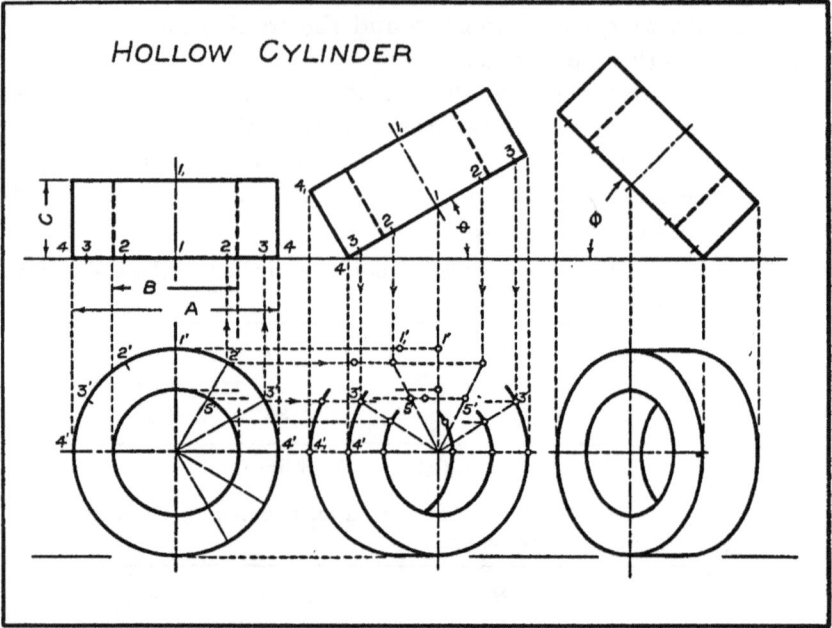

HOLLOW CYLINDER

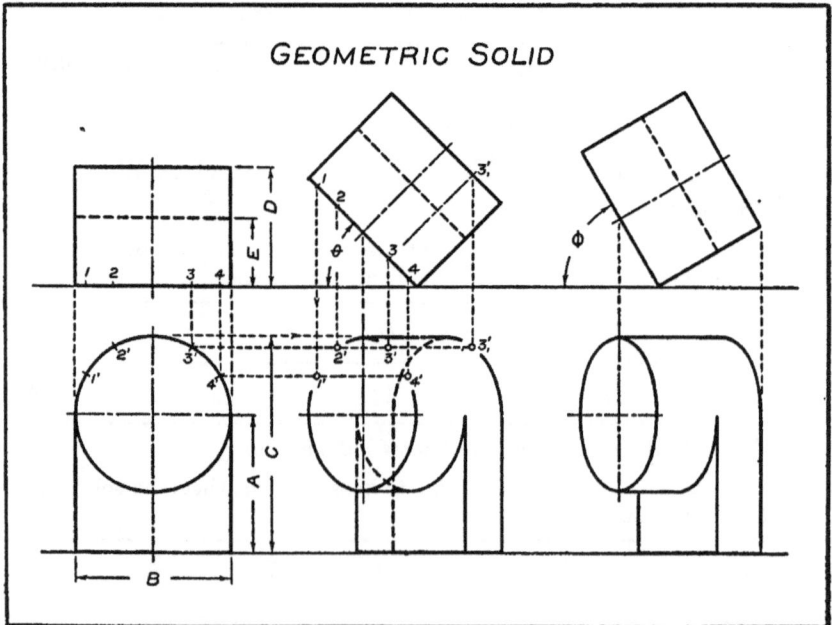

GEOMETRIC SOLID

PROJECTION OF LETTER (P)

This drawing shows the top and the front views of the letter **P** in three positions.

For the first position, draw front and top views as shown. Divide the circular arc, whose center is to be found by trial, into any number of parts, and project the points found to the top view. Transpose the top view to the second and third positions and project the front views.

Problem 1.—Draw similar positions of the letter with the following dimensions:

A	B	C	D	E	F	G	H	θ	ϕ
$3\frac{5}{8}$	1	3	$1\frac{1}{4}$	$\frac{3}{4}$	$\frac{1}{4}$	$2\frac{1}{8}$	$1\frac{1}{4}$	30°	60°

Show all hidden lines in the front view of the second position.

Problem 2.—Draw similar positions of the letter when transformed into the letter **R** with the following dimensions:

A	B	C	D	E	F	G	H	θ	ϕ
$3\frac{1}{2}$	$\frac{7}{8}$	$2\frac{7}{8}$	$1\frac{1}{8}$	$\frac{11}{16}$	$\frac{7}{32}$	$2\frac{1}{8}$	$1\frac{3}{16}$	15°	45°

Show all hidden lines in the front view of the third position.

PROJECTION OF LETTER (S)

This drawing shows the front and the top views of the letter S in three positions.

For the first position, draw front and top views as shown. Assume points on the circular arcs of the front view and project these points to the top view. Transpose the top view to the second and third positions and project the front views.

Problem 1.—Draw the letter in the three positions having the dimensions shown. Let $\theta = 30°$, and $\phi = 45°$.

Problem 2.—For the first position, draw the front view as shown and project the top view when removed one inch from V. For the second position, turn the top view in a counterclockwise direction to an angle of 30° with V. For the third position, turn the top view in a clockwise direction to an angle of 45° with V. Let a point of the letter be in contact with V in the second and the third positions.

PLATE 11

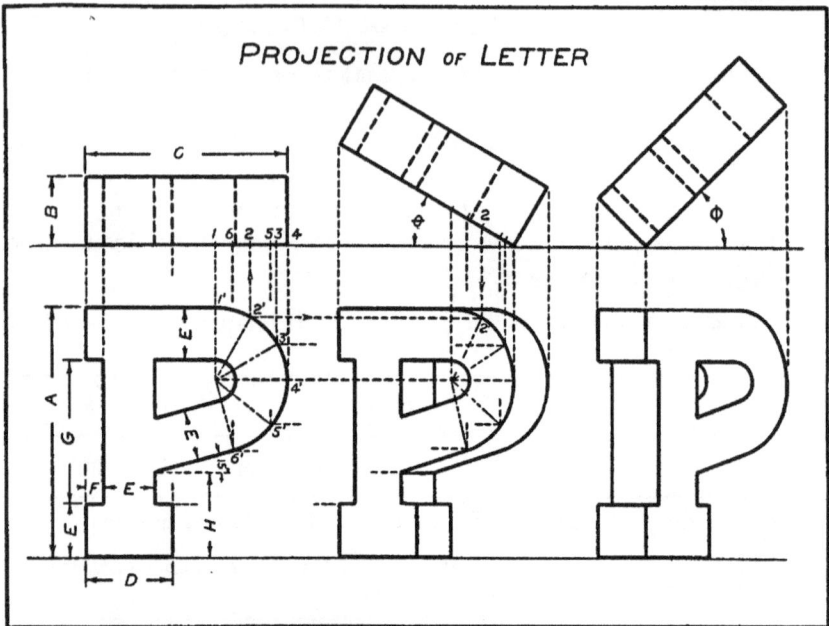

PROJECTION OF LETTER

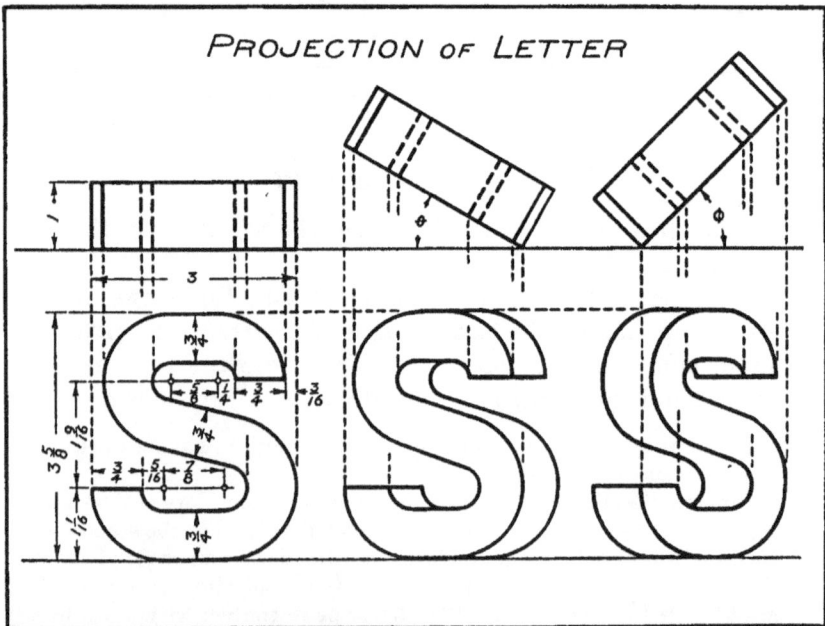

PROJECTION OF LETTER

PLATE 12

ANGLES GREATER THAN 90 DEGREES

Angles greater than 90 degrees, as called for in Problem 2 in each of the six following drawings, are drawn with the triangles. To draw an angle greater than 90 degrees draw its supplement, which is the difference between the required angle and 180 degrees, or straight angle.

PRISMS

To find the projections of a right prism when inclined to the horizontal and the vertical planes of projection.

Draw the top and the front views as shown for the first position. Transpose the front view to the second position, turning it through the required angle, and project the top view. Transpose the top view to the third position, turning it through the required angle, and, finally, find the front view.

TRIANGULAR PRISM
Equilateral Base

Problem 1.—Draw top and front views of the prism in positions similar to those shown. Let $A = 2$ inches, $B = 3$ inches, $\theta = 30°$, and $\phi = 45°$.

Problem 2.—Draw top and front views for the first position, when the prism is turned about its central vertical axis through an angle of 180°. Let $A = 1\frac{1}{4}$ inches and $B = 2\frac{3}{4}$ inches.

Find the front views when the top views are turned about df. Let $\theta = 135°$, $\phi = 150°$, and d be in contact with V for the third position.

PENTAGONAL PRISM
Regular Base

Problem 1.—Draw top and front views of the prism in positions similar to those shown. Let $A = 1\frac{1}{4}$ inches, $B = 3$ inches, $\theta = 60°$, and $\phi = 45°$.

Problem 2.—Draw top and front views for the first position, when the prism is turned about its central vertical axis through an angle of 180°. Let $A = 1\frac{1}{8}$ inches, $B = 2\frac{7}{8}$ inches.

Find the front views when the top views are turned about $c_1 d_1$. Let $\theta = 135°$, $\phi = 150°$, and b be in contact with V for the third position.

PLATE 13

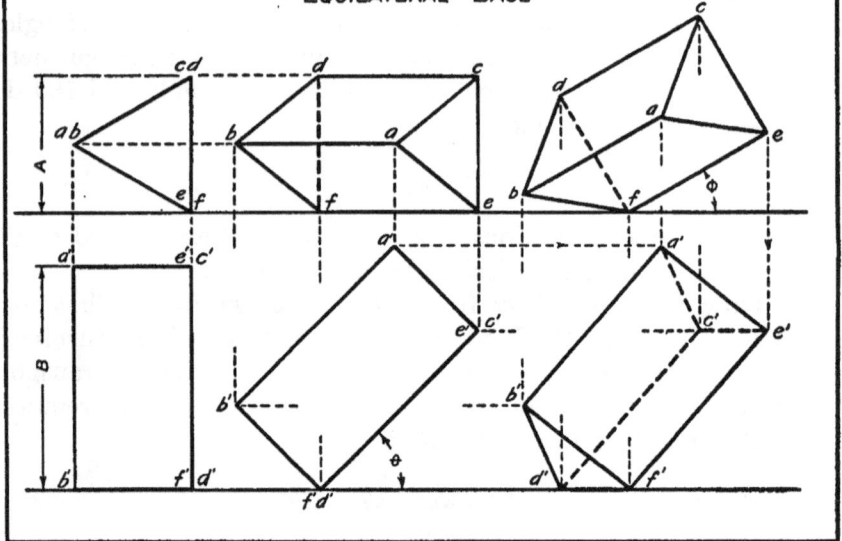

TRIANGULAR PRISM
EQUILATERAL BASE

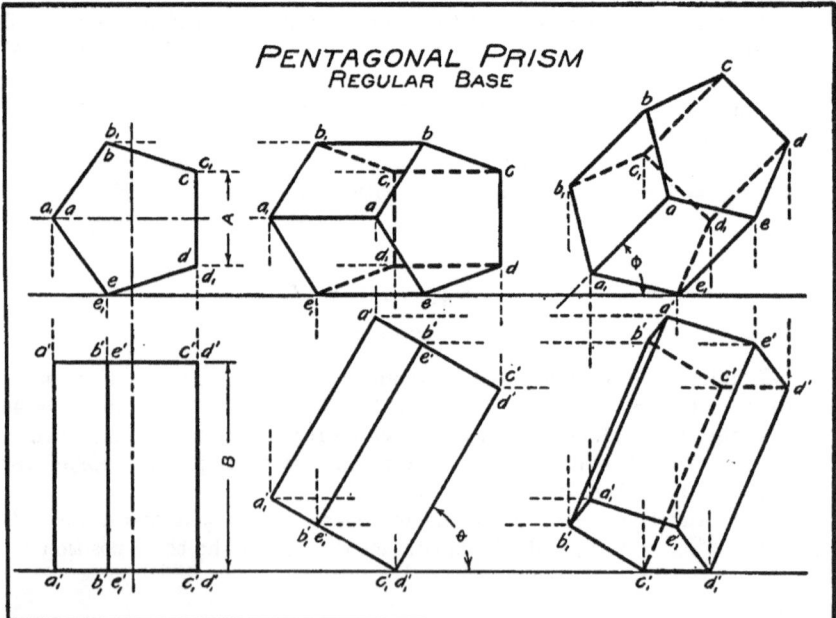

PENTAGONAL PRISM
REGULAR BASE

PLATE 14

PYRAMIDS

To find the projections of a pyramid when resting on one corner of its base, and inclined to the horizontal and the vertical planes.

Draw the top view, making any desired angle with GL, and find the front view. Transpose the front view, turning it through the required angle, and draw the top view. Transpose the top view, turning it through the required angle, and, finally, find the front view.

SQUARE PYRAMID

Problem 1.—Draw top and front views of the pyramid in positions similar to those shown. Let A = 2 inches, B = $3\frac{1}{4}$ inches, α = 20°, θ = 45°, and ϕ = 30°.

Problem 2.—Draw top and front views of the pyramid when inclined to H and V and resting on a corner of its base. Let A = $1\frac{1}{2}$ inches, B = 3 inches, α = 15°, θ = 120°, ϕ = 150°, and let a corner of its base be in contact with V.

HEXAGONAL PYRAMID
Regular Base

Problem 1.—Draw top and front views of the pyramid when inclined to H and V and resting on a corner of its base. Let A = 1 inch, B = $3\frac{1}{4}$ inches, θ = 45°, and ϕ = 60°.

Problem 2.—Draw top view of the pyramid for the first position, when turned counterclockwise about f through an angle of 30°, and project front view. Let A = $1\frac{1}{4}$ inches and B = $3\frac{3}{4}$ inches. Find top and front views when turned about one edge of its base resting on H. Let θ = 120°, ϕ = 135°, and let a corner of the base be in contact with V.

PLATE 15

SQUARE PYRAMID

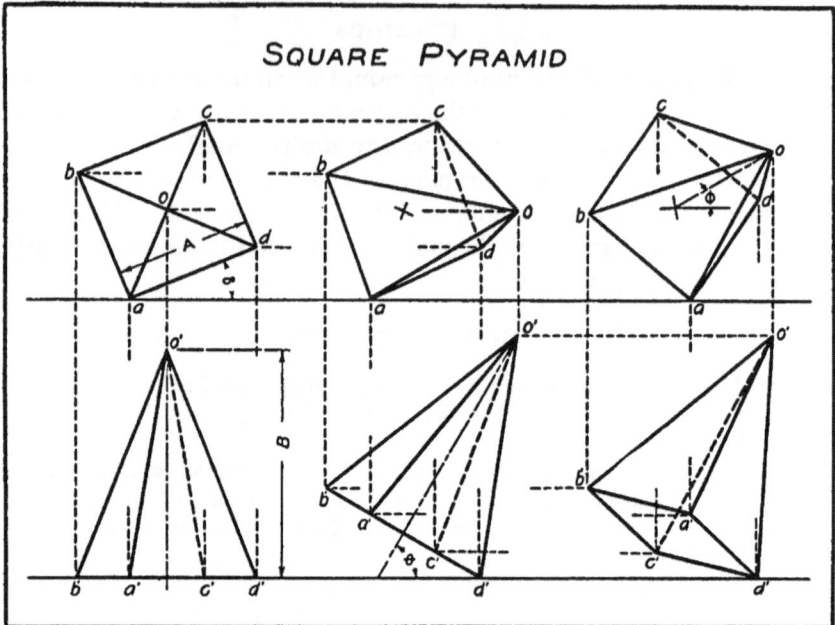

HEXAGONAL PYRAMID
REGULAR BASE

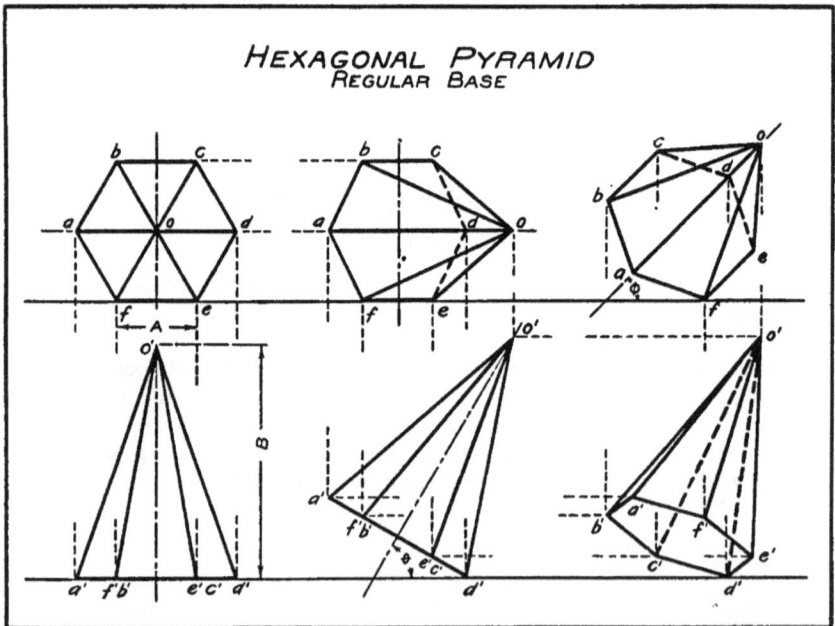

PLATE 16

CONE AND CYLINDER

To find the projections of a cone or a cylinder with its axis inclined to the horizontal and the vertical planes.

Draw the top and the front views of the cylinder or the cone. Divide the top view into equal parts as shown, and project to the front view. Transpose the front view, turning it through a required angle, and project the top view. Transpose the top view, turning it through a required angle, and, finally, project the front view.

RIGHT CONE

Problem 1.—Draw top and front views of a right cone when resting on a point of its base and inclined to H and V. Let $A = 2$ inches, $B = 3\frac{1}{2}$ inches, $\theta = 45°$, and $\phi = 45°$.

Problem 2.—Draw top and front views of a right cone when resting on a point of its base and inclined to H and V. Let $A = 1\frac{3}{4}$ inches, $B = 3\frac{1}{4}$ inches, $\theta = 135°$, and $\phi = 135°$.

RIGHT CYLINDER

Problem 1.—Draw top and front views of a right cylinder when resting on a point of its base and inclined to H and V. Let $A = 2$ inches, $B = 3$ inches, $\theta = 45°$, and $\phi = 45°$.

Problem 2.—Draw top and front views of a right cylinder when resting on a point of its base and inclined to H and V. Let $A = 1\frac{3}{4}$ inches, $B = 2\frac{1}{4}$ inches, $\theta = 135°$, and $\phi = 135°$.

PLATE 17

RIGHT CONE

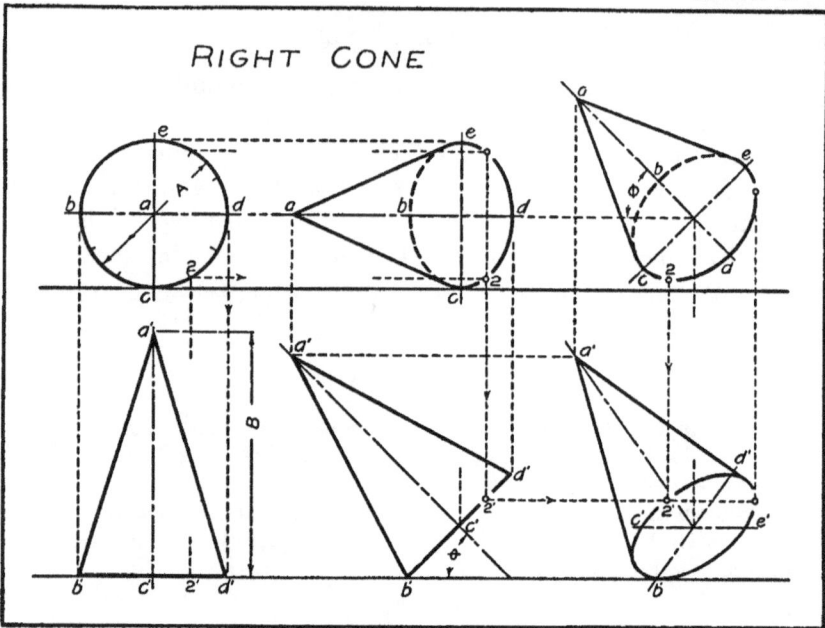

RIGHT CYLINDER

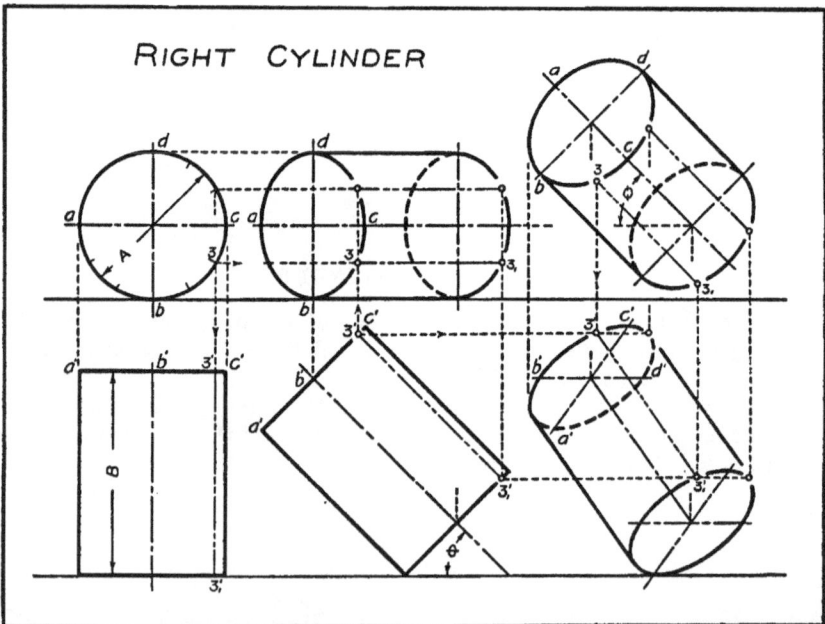

PLATE 18

SQUARE PRISMS

To draw the projections of a square prism resting on the edge of another square prism.

Draw the front view as shown in the first position and project the top view. Transpose the top view to the second position, turning it through a required angle. The front view of the second position is then found by projecting from the top view of the second position and the front view of the first position.

Problem 1.—Find the front view of the two prisms when $\theta = 30°$ and $\phi = 45°$.

Problem 2.—Find the front view of the two prisms when the top view of the second position is turned in a clockwise direction through an angle of 30°, instead of counterclockwise as shown. Let $\theta = 30°$.

CROSS AND PRISM

To draw the projections of a cross and a prism with one edge of the former resting on the latter.

Draw the front view, as shown in the first position, and project the top view, including all hidden lines, and proceed as explained for Square Prisms.

Problem 1.—Find the front view of the cross and the prism when $\theta = 60°$ and $\phi = 30°$. Determine the height of the prism by measurement.

Problem 2.—Find the front view of the cross and the prism when the top view of the second position is turned in a counterclockwise direction through an angle of 45°, instead of clockwise as shown. Let $\theta = 75°$. Determine the height of the prism by measurement.

PLATE 19

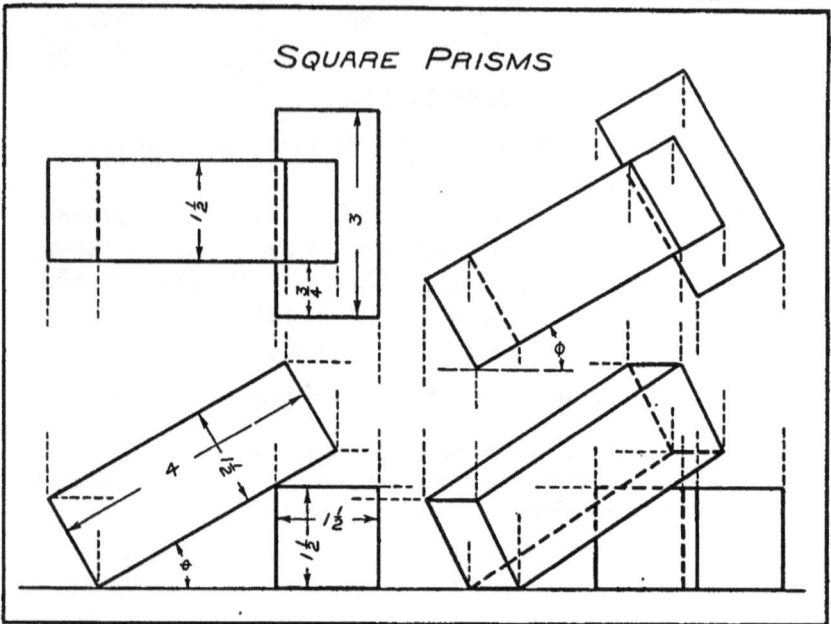

SQUARE PRISMS

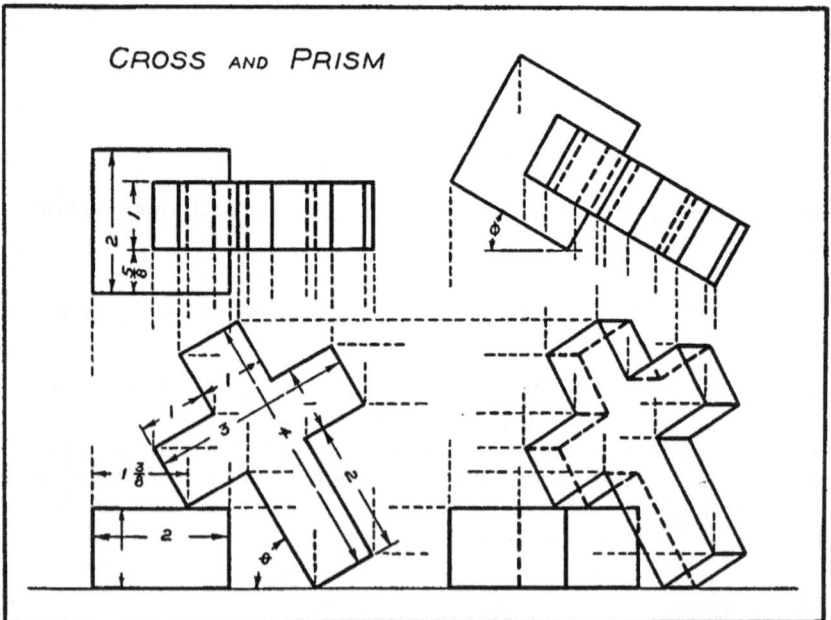

CROSS AND PRISM

PLATE 20

LETTER AND PRISM (Z)

To draw the projections of a letter resting on an edge of a triangular prism.

Draw the front view and project the top view, showing all hidden lines. If the letter contains curved lines it will be necessary to assume a number of points in the front view and project these points to the top view, as shown by points 1 and 2 in the letter R. Transpose the top view to the second position, turning it through a required angle, and find the front view.

Problem 1.—Design the letter H, of the same general dimensions and position as shown for the letter Z, and find the front view when both letter and prism are inclined to V. Let $\theta = 30°$ and $\phi = 60°$.

Problem 2.—Draw the front view of the letter shown when moved over the prism toward the left, until the left edge rests on the auxiliary horizontal plane, and the right end rests on the prism, and when both are inclined to V. Let the bottom of the letter be inclined 30° to H, and $\phi = 120°$.

LETTER AND PRISM (R)

Problem 1.—Find the front view of the letter and the prism when the top view of the second position is turned in a counterclockwise direction, instead of clockwise as shown. Let $\theta = 30°$ and $\phi = 120°$. The height of the prism is to be found by measurement.

Problem 2.—Find the front view when the letter is moved over the prism toward the right, until the right edge rests on the auxiliary horizontal plane, and the left end rests on the prism, and when both are inclined to V. Let the bottom of the letter be inclined 30° to H, and $\phi = 60°$. The height of the prism is to be found by measurement.

PLATE 21

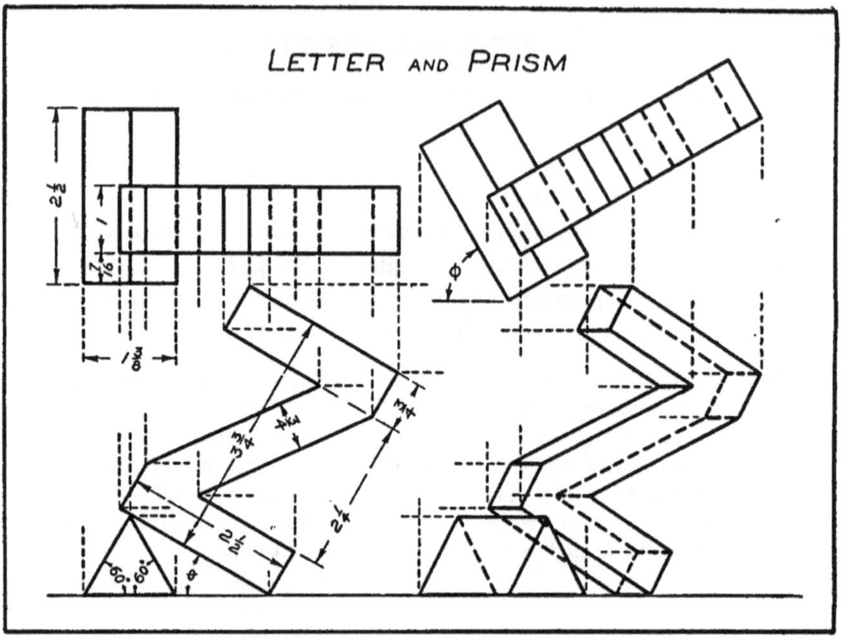

LETTER AND PRISM

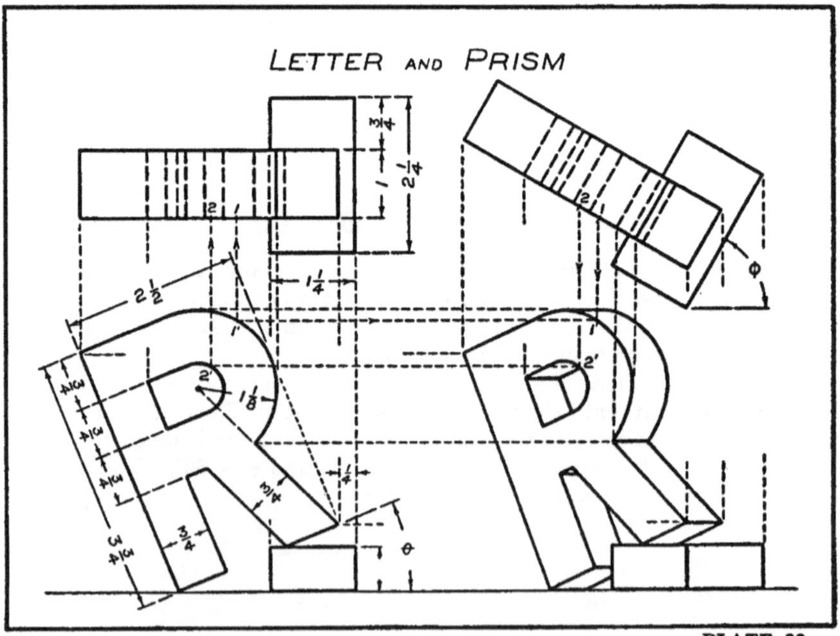

LETTER AND PRISM

PLATE 22

CYLINDER AND PRISM

To draw the projections of an inclined cylinder resting on an edge of a triangular prism.

Draw the front view and project the top view. Transpose the top view to the second position, turning it through a required angle, and project the front view.

The ends of the cylinder in the top view of the first position are found by dividing the surface of the cylinder, by aid of auxiliary half-circles, in both views, into a number of equal parts as shown.

Problem 1.—Find the front view of the cylinder and the prism when both are inclined to V and of the following dimensions:

A	B	C	D	E	θ	ϕ
$3\frac{1}{8}$	$2\frac{1}{8}$	$1\frac{3}{4}$	$1\frac{1}{2}$	$2\frac{1}{2}$	45°	15°

Omit hidden lines in the second position.

Problem 2.—Find the front view of the cylinder and prism when both are inclined to V and of the following dimensions:

A	B	C	D	E	θ	ϕ
$3\frac{1}{4}$	$2\frac{1}{4}$	$1\frac{3}{4}$	$1\frac{1}{2}$	$2\frac{1}{2}$	135°	30°

Cylinder will be on the right side of the prism instead of on the left as shown. Show all hidden lines.

CONE AND CYLINDER

Problem 1.—Find the front view of the cone and the cylinder when both are inclined to V and of the following dimensions:

A	B	C	D	θ	ϕ
$4\frac{1}{4}$	$2\frac{1}{8}$	$2\frac{1}{8}$	$2\frac{1}{2}$	45°	30°

Show all hidden lines.

Problem 2.—Find the front view of the cone and the cylinder when both are inclined to V and of the following dimensions:

A	B	C	D	θ	ϕ
$4\frac{1}{2}$	$2\frac{1}{4}$	$2\frac{1}{8}$	$2\frac{1}{2}$	135°	30°

Cone will be on the left side of the cylinder instead of on the right as shown. Show all hidden lines.

PLATE 23

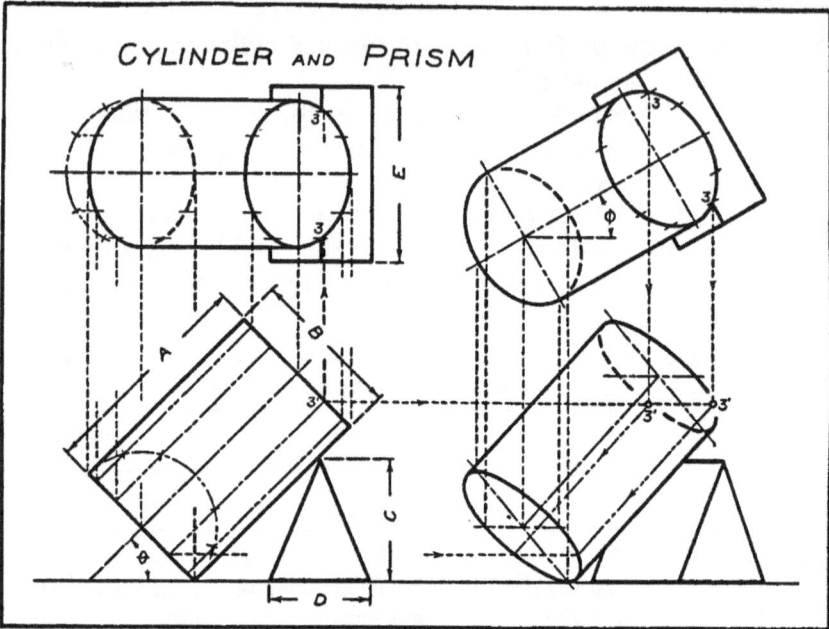

CYLINDER AND PRISM

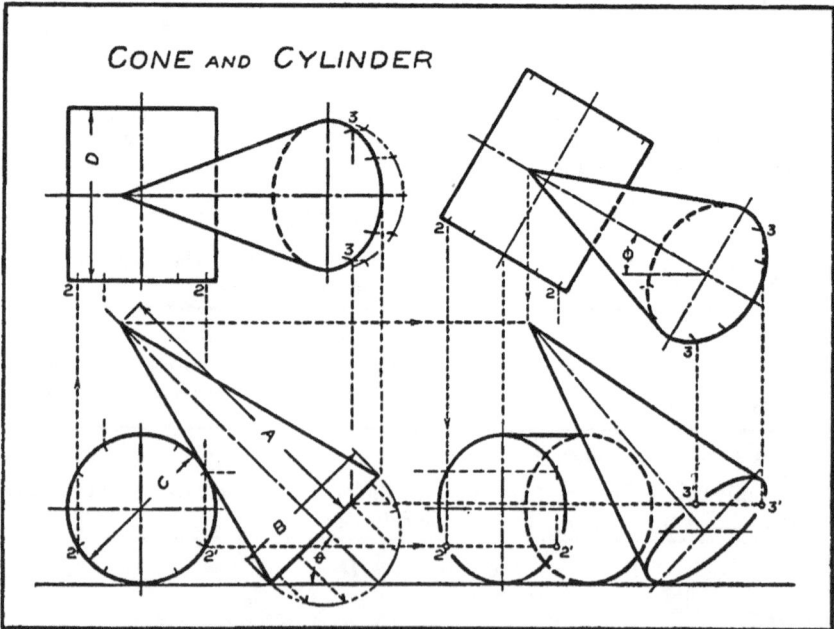

CONE AND CYLINDER

PLATE 24

PROJECTIONS ON THREE PLANES

The four following drawings show the top view, the front view, and the side view of simple objects.   See page 23.

The side view of an object is generally shown on the right side of the front view.   It may, however, be shown on the left side, if some special features of the object warrant the change.

PROJECTION PROBLEMS
(First Drawing)

Problem 1.—Draw three views of the first figure when turned about its central vertical axis from the position shown, through an angle of 180°.

Draw three views of the second figure when turned about its central vertical axis from the position shown, through an angle of 180°.   Let $\theta = 30°$.

Assume distances from H, V, and P, for both figures.

Problem 2.—Draw three views of the first figure when turned through an angle of 90° about its central vertical axis in a clockwise direction.

Draw three views of the second figure when $\theta = 135°$.

Assume distances from H, V, and P, for both figures.

PROJECTION PROBLEMS
(Second Drawing)

Problem 1.—Draw three views of the first figure in the position shown, omitting the rectangular hole.

Draw three views of the second figure, including the hole, when $\theta = 60°$.

Problem 2.—Draw three views of the first figure when turned about its central vertical axis through an angle of 90°.

Draw three views of the second figure when $\theta = 135°$.

PLATE 25

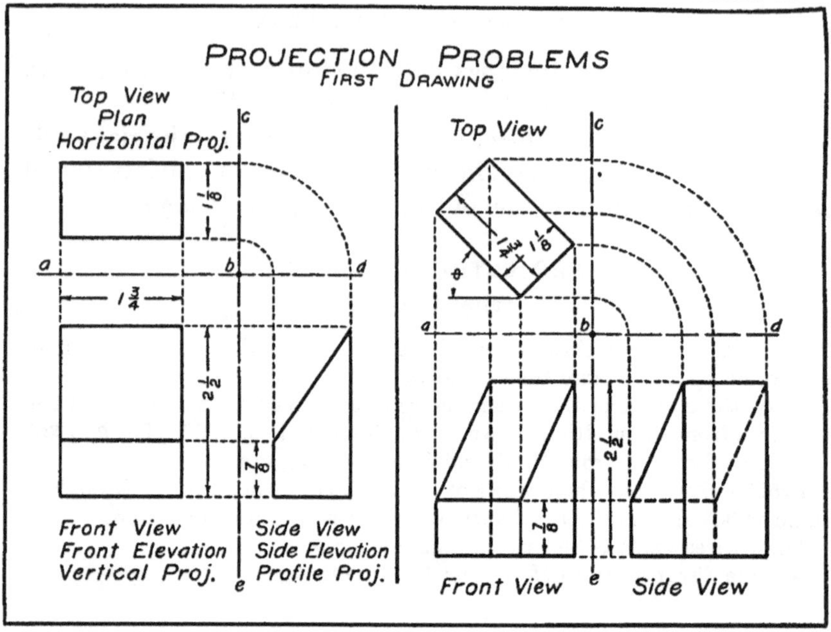

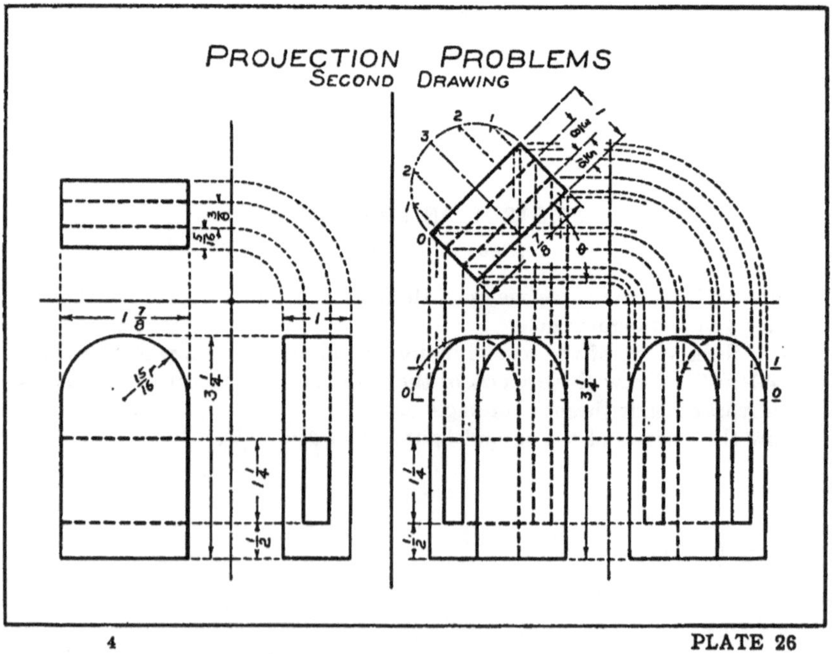

PROJECTION PROBLEMS
FIRST DRAWING

Top View
Plan
Horizontal Proj.

Top View

Front View
Front Elevation
Vertical Proj.

Side View
Side Elevation
Profile Proj.

Front View

Side View

PROJECTION PROBLEMS
SECOND DRAWING

PLATE 26

PROJECTIONS ON THREE PLANES
(Continued)

In many cases a figure may be completely shown by two views, but if three views will show its construction more clearly, then three views are given. The figures in the drawings on page 49 are completely shown by the front and the side views alone, but the addition of the top views gives a greater degree of clearness to their form.

A close study of the figures of the third and fourth drawings of PROJECTION PROBLEMS shows that two views of any one figure are insufficient for its complete representation.

PROJECTION PROBLEMS
(Third Drawing)

Problem 1.—Draw three views of the first figure when removed ⅜ inch from V, ⅞ inch from H, and ¼ inch from P.

Draw three views of the second figure when removed ⅝ inch from V, ⅞ inch from H, and ¼ inch from P.

Problem 2.—Draw three views of the first figure when turned through an angle of 180° about its vertical axis. Let the figure be removed ¼ inch from V, ¾ inch from H, and ¼ inch from P.

Draw three views of the second figure when turned through an angle of 90° about its vertical axis. Let the figure be removed ¼ inch from V, ¾ inch from H, and ⅜ inch from P.

PROJECTION PROBLEMS
(Fourth Drawing)

Problem 1.—Draw three views of the first figure when removed $\frac{5}{16}$, ¾, and ⅝ inches from V, H, and P, respectively.

Draw three views of the second figure having its central vertical axis 1¼ inches from both V and P. Let the figure rest on an auxiliary H plane which is 4¾ inches from H.

Problem 2.—Draw three views of the first figure when turned about its central vertical axis through a half-revolution. Let the axis be ⅞ inch from V, 1¼ inches from P, and let the top of the figure be ⅝ inch below H.

Draw three views of the second figure when turned about its central vertical axis through a quarter-revolution in a counterclockwise direction. Let the axis be 1½ inches from V, 1¼ inches from P, and let the top of the figure be ⅞ inch below H.

PLATE 27

PROJECTION PROBLEMS
THIRD DRAWING

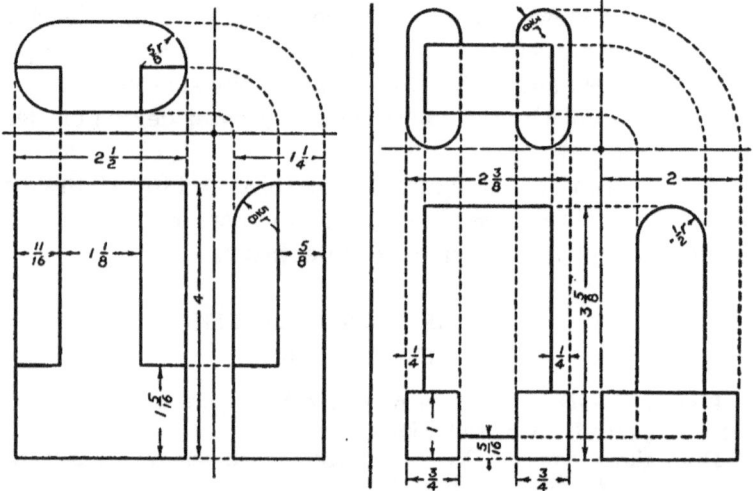

PROJECTION PROBLEMS
FOURTH DRAWING

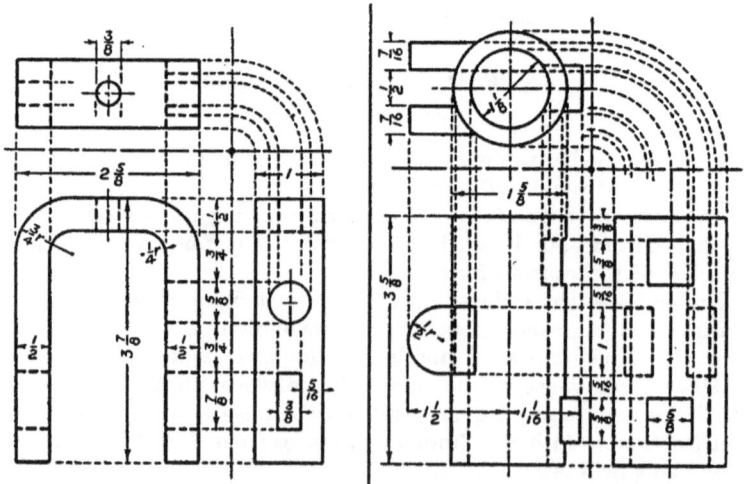

PLATE 28

PROJECTION PROBLEMS

(Fifth Drawing)

See page 23

Problem 1.—Draw front, top, and auxiliary views with the following dimensions:

	A	*B*	*C*	*D*	*θ*
First Figure.........	$2\frac{1}{2}$	$1\frac{5}{8}$	$1\frac{1}{4}$	$\frac{1}{2}$	30°
Second Figure.......	$2\frac{1}{2}$	1	$\frac{1}{2}$		60°

Problem 2.—Draw front, top, and auxiliary views with the following dimensions:

	A	*B*	*C*	*D*	*θ*
First Figure.........	$2\frac{3}{8}$	$1\frac{1}{2}$	$1\frac{1}{4}$	$\frac{1}{8}$	30°
Second Figure.......	$2\frac{3}{8}$	$1\frac{1}{2}$	0		45°

PROJECTION PROBLEMS

(Sixth Drawing)

Problem 1.—Draw four views of each object in positions similar to those shown and of the following dimensions:

	A	*B*	*C*	*θ*
First Figure..........	$2\frac{1}{2}$	$1\frac{3}{4}$	$1\frac{3}{4}$	60°
Second Figure........	$3\frac{1}{2}$	1	2	45°

Show all hidden lines in both objects.

Problem 2.—Draw four views of each of the two objects using the following dimensions:

	A	*B*	*C*	*θ*
First Figure..........	$2\frac{1}{2}$	$1\frac{3}{4}$	$1\frac{1}{2}$	45°
Second Figure........	$3\frac{1}{2}$	1	2	45°

The first object is to be turned about its central vertical axis through an angle of 180°; the second, to be turned about its central vertical axis in a clockwise direction through an angle of 90°. See second paragraph page 48.

Show all hidden lines in both objects.

PROJECTION PROBLEMS
FIFTH DRAWING

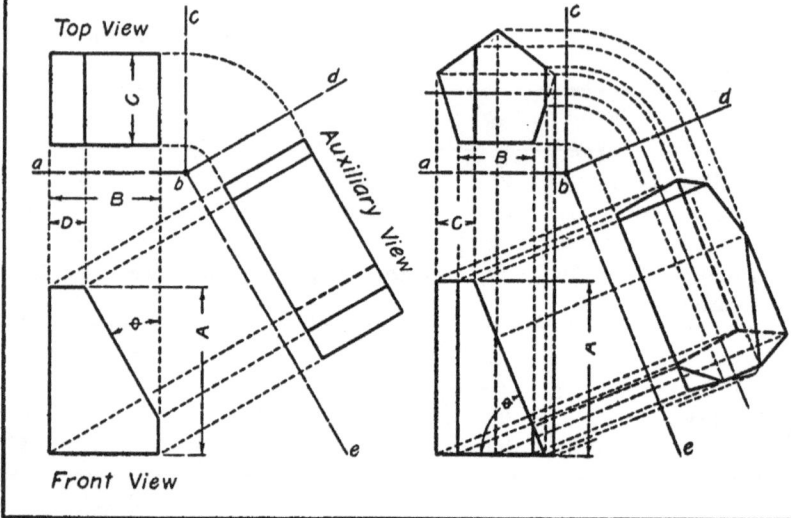

Top View

Front View

Auxiliary View

PROJECTION PROBLEMS
SIXTH DRAWING

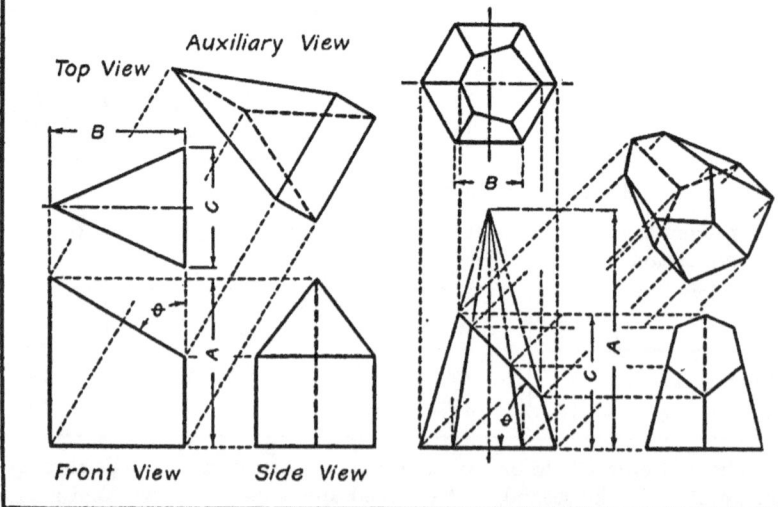

Auxiliary View

Top View

Front View   Side View

PLATE 30

SECTION II

DEVELOPMENTS AND INTERSECTIONS

DEVELOPMENT

The shape of the surface of an object when laid out on a plane is its development, or pattern.

In the developments of the following drawings, the overall dimensions, a number of which are not shown, although the dimension lines are drawn, are to be found by accurately measuring the drawing.

RECTANGULAR PRISM

To find the development of a rectangular prism, first draw top and front views of the object. Then draw a horizontal line of indefinite length and lay off distances equal to the lengths of the lateral faces as found from the top view. At these points erect vertical lines, and measure off the lengths of the lateral edges as found from the front view. Complete the lateral area, or surface, by drawing a line connecting the heights of the edges.

Problem 1.—Draw top and front views, and find the development of a prism having a base of $1\frac{1}{4} \times 1\frac{3}{4}$ inches and a height of $2\frac{1}{4}$ inches.

Problem 2.—Draw top and front views, and find the development of a prism having a base of $1\frac{3}{4} \times 2\frac{1}{4}$ inches and a height of $2\frac{1}{4}$ inches. In the projections show the narrow faces of the prism parallel to V.

REVOLVED SURFACE

A revolved surface is a projection which shows the true shape of the face of an object which is not parallel to the regular planes of projection.

TRUNCATED RECTANGULAR PRISM

To find the development of a truncated rectangular prism, first draw top and front views and the revolved surface of the object. Then draw the lateral surface, the length being equal to the perimeter of the top view, and the several heights being equal to the lateral edges found from the front view.

Problem 1.—Draw two views, the revolved surface, and find the development of a truncated prism having a $1\frac{1}{2}$-inch square base and a height of 3 inches. Let the angle in the front view be 30° with a horizontal.

Problem 2.—Draw two views, the revolved surface, and find the development of a truncated prism having a base of $1\frac{1}{4} \times 2\frac{1}{4}$ inches and a height of $2\frac{1}{4}$ inches. In the projections show the narrow faces parallel to V. Assume suitable angle for the inclined surface.

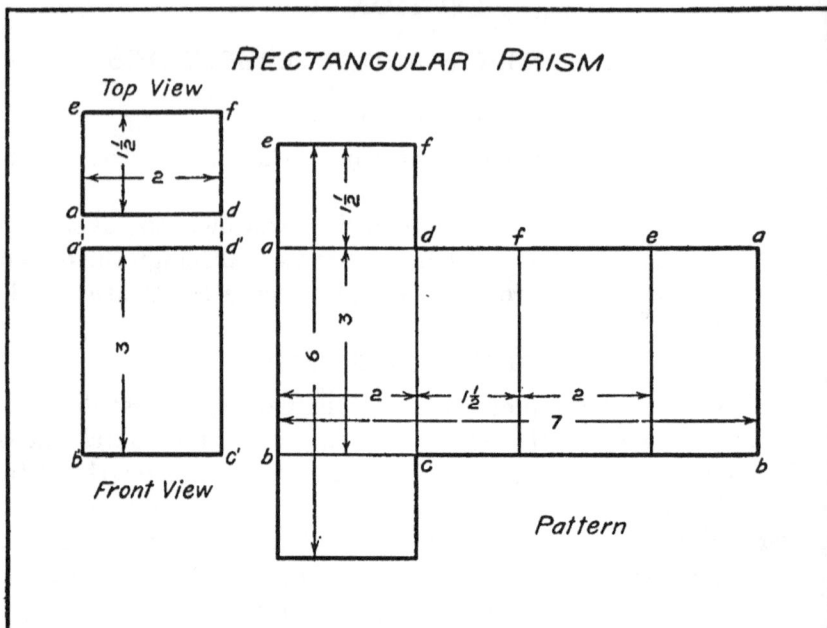

RECTANGULAR PRISM

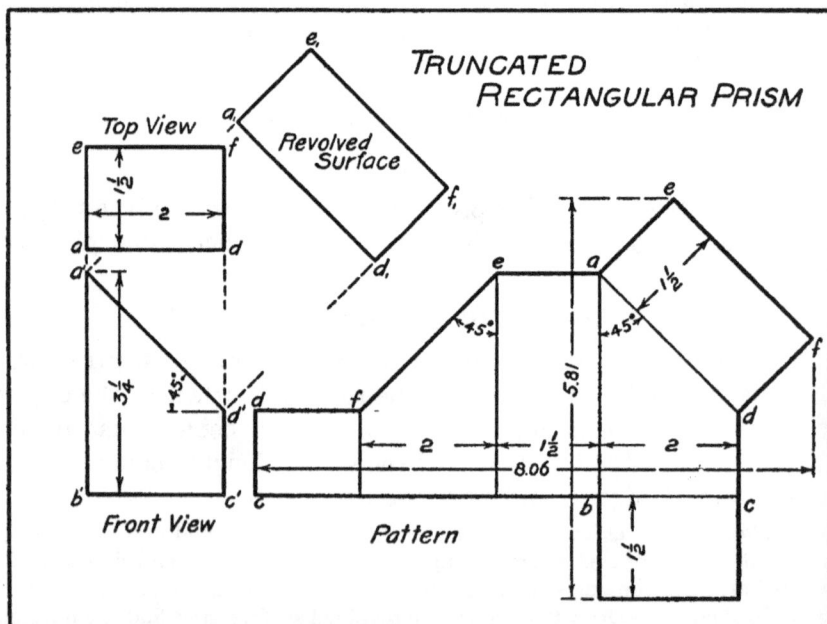

TRUNCATED RECTANGULAR PRISM

PLATE 32

TRIANGULAR WEDGE

To find the development of a triangular object, like the wedge shown in the drawing, first draw top and front views. Lay out a rectangular surface whose height equals the height of the wedge and whose length equals the perimeter of the top view. To this rectangular surface attach triangular surfaces equal to the top and the bottom of the wedge.

The true lengths of the sides of the wedge may be found by calculation, or by measurement of the top view.

Problem 1.—Draw two views, and the developed surface of a triangular wedge. Reduce the horizontal dimensions, shown in the drawing, ¼ inch and increase the vertical dimension ⅛ inch.

Problem 2.—Draw top and front views of the triangular wedge, shown in the drawing, when turned about its central vertical axis through a half revolution. Develop the surface when opened on the edge ad, instead of be as shown.

TRUNCATED SQUARE PRISM

To find the development of a truncated prism similar to that shown, first draw top and front views and the revolved surface. Lay out the lateral surface, the length being found from the top view and the several heights being found from the front view. Attach surfaces equal to the revolved and the bottom surfaces.

Problem 1.—Draw top and front views, the revolved surface, and the development of the prism, when the 60° angle shown in the front view is changed to 45°.

Problem 2.—Draw two views, and the revolved surface, as shown in the drawing, and find the development when opened on the line ab.

PLATE 33

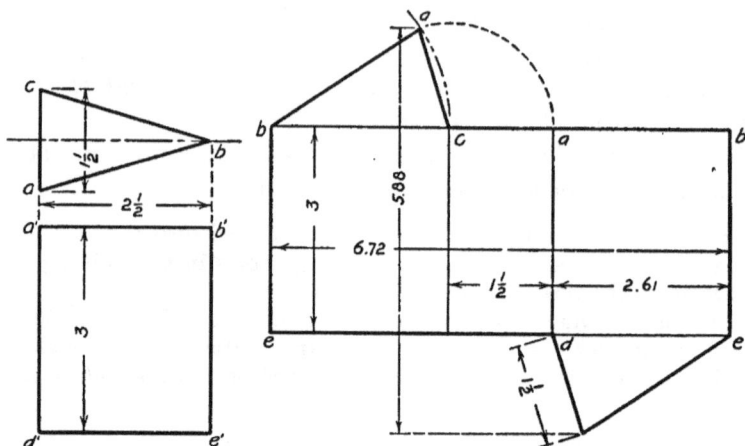

TRIANGULAR WEDGE

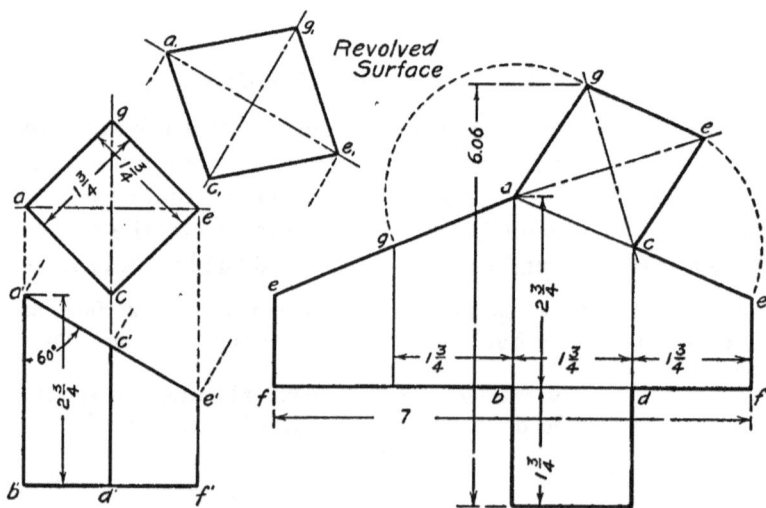

TRUNCATED SQUARE PRISM

PLATE 34

⌊TRUNCATED TRIANGULAR PRISM

To find the development of a triangular prism similar to that shown, draw top and front views, and the revolved surface. Lay out the lateral surface. Attach triangular surfaces equal to the revolved and the bottom surfaces.

Problem 1.—Draw top and front views, the revolved surface, and the development when the 45° angle in the front view is changed to 60°. Show development when opened on the shortest edge of the prism.

Problem 2.—Draw the top view, when revolved through an angle of 180° about its center, and project the front view. Let the front view of the inclined surface be as shown. Project the revolved surface and find development when opened on one of the shortest edges of the prism.

TRUNCATED HEXAGONAL PRISM

To find the development of an hexagonal prism similar to that shown, draw top and front views, and the revolved surface; then proceed as explained for the triangular prism.

Problem 1.—Draw top and front views of the prism showing the inclined surface beginning at a', in the front view, and making an angle of 30° with the horizontal. Project the revolved surface and find the development when opened on the shortest edge of the prism.

Problem 2.—Draw top and front views of the prism when turned about its central vertical axis through an angle of 30°. Let the inclined surface in the front view begin at the axis and make an angle of 45°. Project the revolved surface and find the development when opened on one of the shortest edges of the prism.

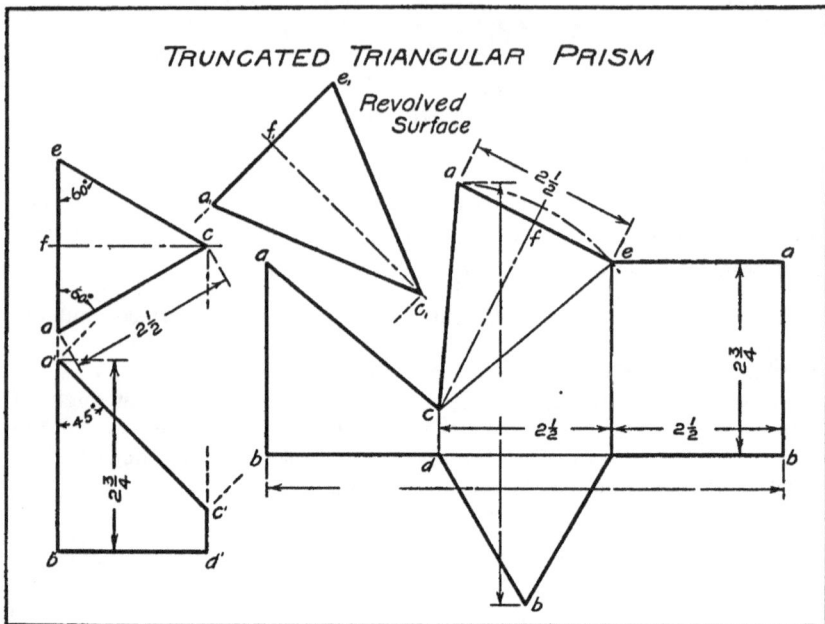

TRUNCATED TRIANGULAR PRISM

Revolved Surface

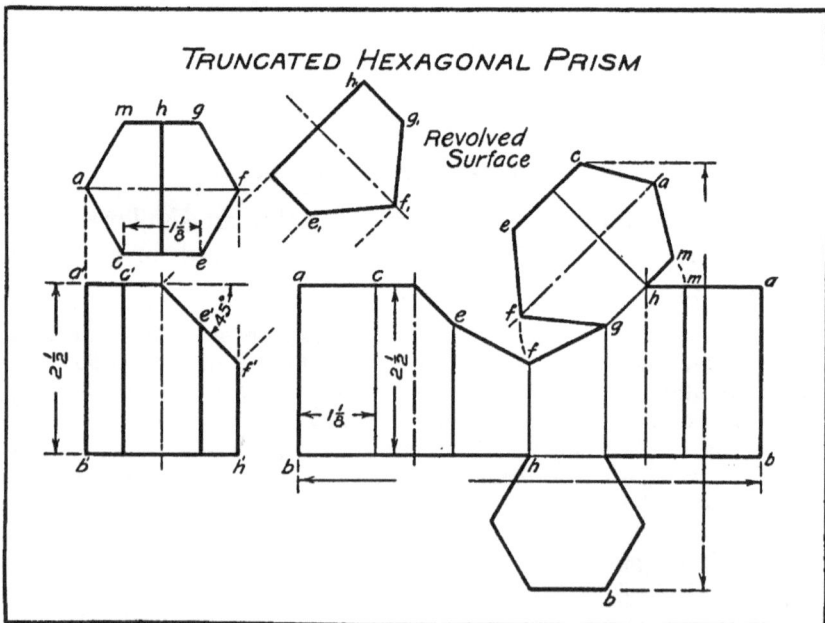

TRUNCATED HEXAGONAL PRISM

Revolved Surface

PLATE 36

TRUNCATED PENTAGONAL PRISM

To find the development of a pentagonal prism similar to that shown, draw the top view and project the front view, then draw the revolved surface. Lay out the lateral surface whose dimensions are found from the top and front views. Attach two plane figures equal to the bottom and the revolved surfaces.

Problem 1.—Draw top and front views of the prism when revolved counterclockwise about its central vertical axis until a face is parallel to V. Show the inclined surface beginning at the axis and making an angle of 30° with the horizontal. Project the revolved surface and find the development when opened on its shortest edge.

Problem 2.—Draw top and front views of the prism when revolved about its central vertical axis through a half-revolution. Let the front view of the inclined surface be as shown. Project the revolved surface and find the development when opened on its longest edge.

TRIANGULAR PYRAMID

Since none of the inclined edges of the pyramid are parallel to either H or V, it is necessary to revolve one edge until it is parallel to one plane. The method shown is as follows: Revolve oc to oc_1, about o as center, until it is parallel to V. Project c_1 to c_1', and draw $o'c_1'$, which will be the true length of the edge.

Problem 1.—Draw top and front views of the pyramid when revolved about its vertical axis through an angle of 180°, and find the development.

Problem 2.—Draw top, front, and side views of a pyramid $3\frac{3}{4}$ inches high having a triangular base, the length of each side being $2\frac{1}{4}$ inches. Let one of its lateral edges be parallel to P and visible on V, and find the development.

For additional figures see page 90.

TRUNCATED PENTAGONAL PRISM

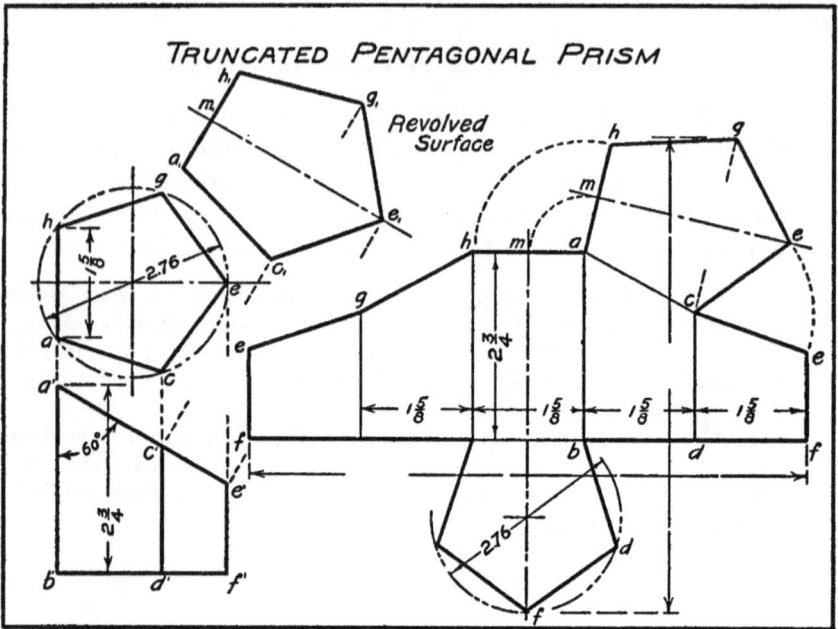

Revolved Surface

TRIANGULAR PYRAMID

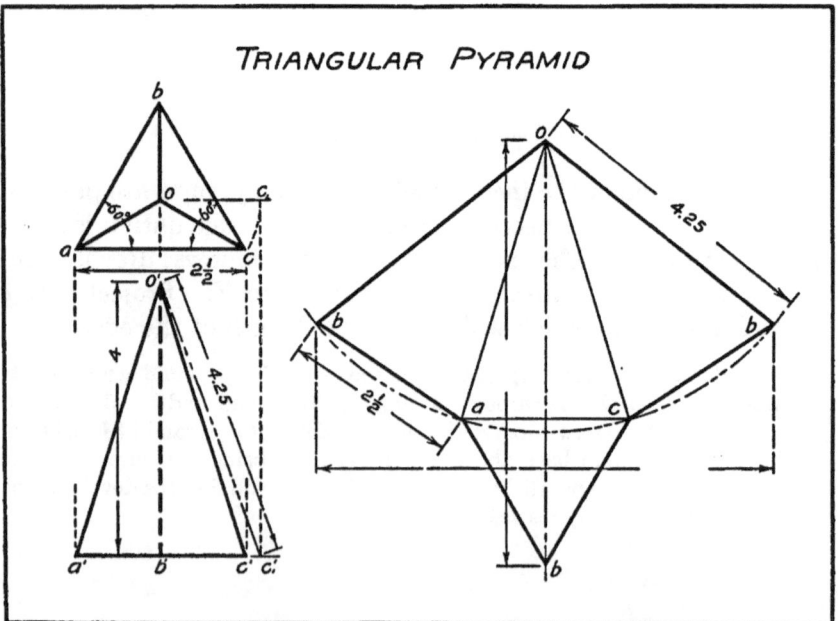

PLATE 38

TRUNCATED SQUARE PYRAMID

To find the true lengths of the edges in this pyramid, it is necessary to revolve one edge into a position parallel to V. The drawing shows the revolved position of one edge in which the true lengths are shown by $f'_1 b'_1$, and $e'_1 b'_1$.

Problem 1.—Draw top and front views, also the revolved surface, and find the development of the truncated prism when opened on one of its short edges.

Problem 2.—Draw top, front, and side views of the truncated pyramid, also the revolved surface, and find the development when the inclined surface makes an angle of 30° with the horizontal plane. Let the height be $2\frac{1}{4}$ inches and let the pattern be opened on one of its shortest edges.

For additional figures see page 91.

SCALENE PYRAMID

Revolve two edges into positions parallel to V. Their true lengths are then shown by $o'b'_1$, and $o'c'_1$.

Problem 1.—Draw the base of the pyramid when turned through an angle of 180° from the position shown, the apex to remain as in the drawing. Complete the top view, and project the front view. Find the development when opened on one of its long edges.

Problem 2.—Draw top and front views of the pyramid as shown, and let it be truncated by a horizontal plane $1\frac{3}{4}$ inches from its base. Find the development when opened on the long edge.

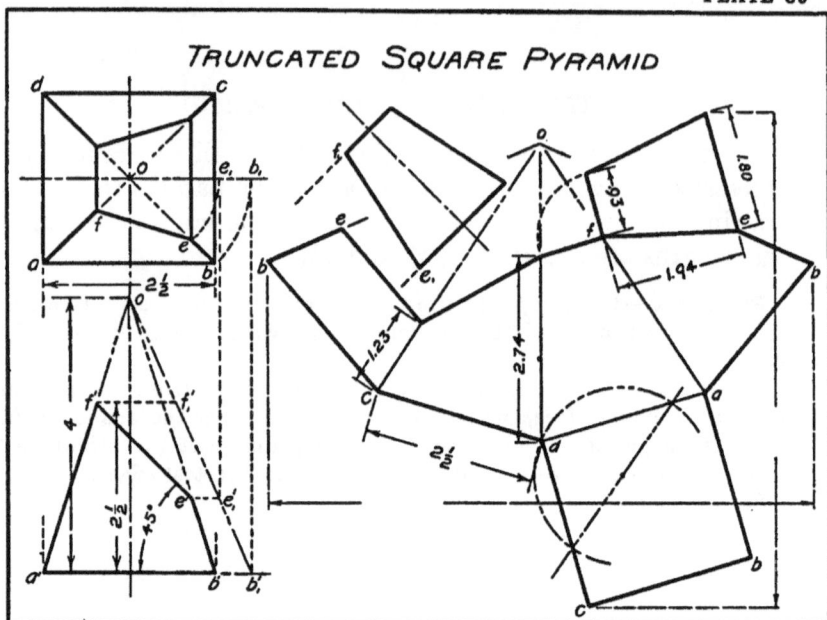

TRUNCATED SQUARE PYRAMID

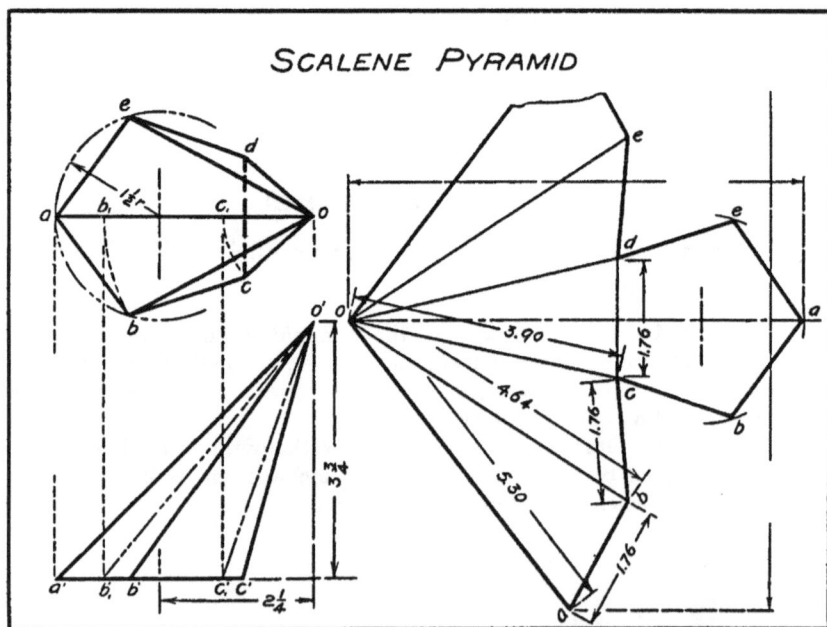

SCALENE PYRAMID

PLATE 40

TRUNCATED CYLINDER AND CONE

To find the revolved surface of a truncated cylinder or a cone, the top view is divided into any convenient number of equal parts. These divisions are projected to the front view. The revolved surface, which is an ellipse, may then be found by drawing lines at right angles to the surface from the points in the front view, and on these laying off the various widths as found in the top view. A study of point 2 in the various views will make the method clear.

TRUNCATED CYLINDER

Problem 1.—Draw top and front views, the revolved surface, the side view of the inclined surface, and find the development of the truncated cylinder when opened on its shortest element. Let $A = 2$ inches, $B = 3\frac{1}{2}$ inches, and $C = 1$ inch.

Problem 2.—Draw top and front views, the revolved surface, the side view of the inclined surface, and find the development of the truncated cylinder when opened on its longest element. Let $A = 1\frac{3}{4}$ inches, $B = 3\frac{1}{4}$ inches, and let the inclined surface make an angle of 60° with the horizontal plane.

TRUNCATED CONE

The true lengths of the elements of the cone may be found by revolving them about the axis into positions parallel to V.

Problem 1.—Draw top, front, and side views, the revolved surface, and find the development of the truncated cone when opened on its shortest element. Let $A = 2\frac{1}{2}$ inches, $B = 4$ inches, $C = 1$ inch, and $\theta = 45°$.

Problem 2.—Draw top, front, and side views, the revolved surface, and find the development of the truncated cone when opened on its longest element. Let $A = 2\frac{3}{4}$ inches, $B = 4$ inches, $C = 1$ inch, and $\theta = 60°$.

PLATE 41

TRUNCATED CYLINDER

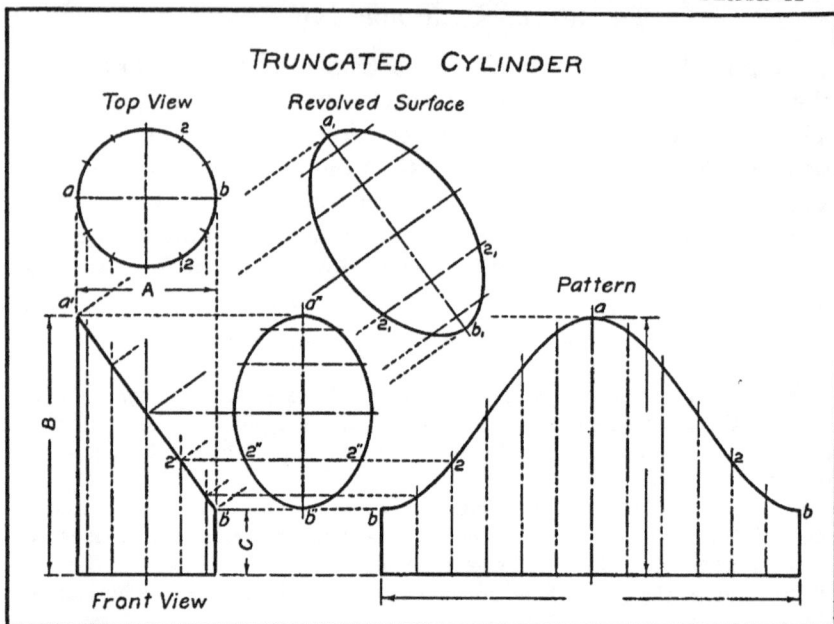

Top View   Revolved Surface   Pattern   Front View

TRUNCATED CONE

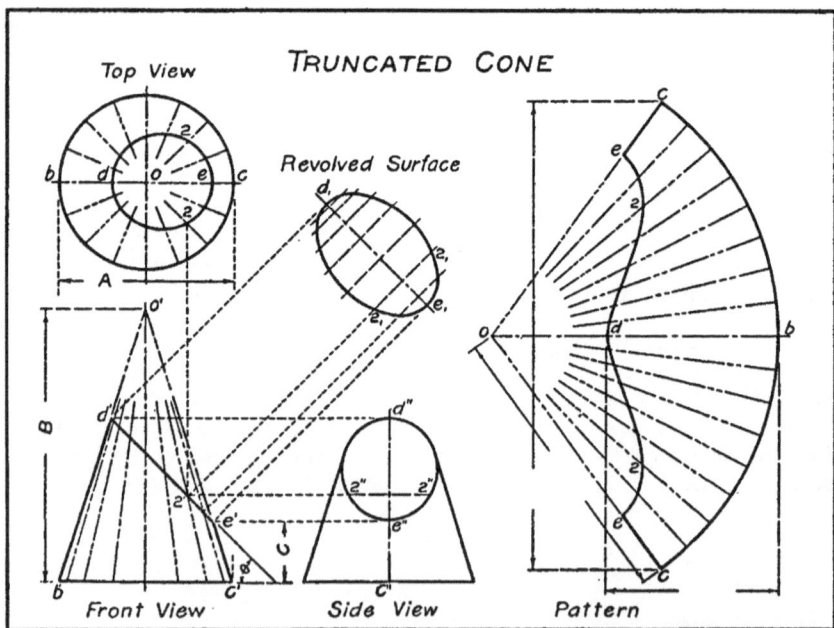

Top View   Revolved Surface   Front View   Side View   Pattern

PLATE 42

Conic Sections
(First and Second Drawings)

The intersection formed by a cone and a plane, making the same angle with the axis as the elements, is a parabola.

The intersection formed by a cone and a plane, making a smaller angle with the axis than the elements, is an hyperbola.

The curves shown in the first drawing are found by the method of intersecting elements. The curves shown in the second drawing are found by the method of cutting planes.

Draw two views showing a cone cut by a plane. Draw V and H projections of an element intersected by a plane giving c' and c as shown in the first drawing; or, draw V and H projections of a horizontal cutting plane giving c' and c as shown in the second drawing. The point formed by either of these methods will be one point on the required curve. The location of the point in the revolved view, which will give the true shape of the curve, is found by measurement from the top or the end views. Other points are found similarly.

To find the development, draw a sector the length of whose arc equals the circumference of the base of the cone, and draw elements the location of which are found from the top view, as in the first drawing; or, draw circular arcs the location of which are found from the front view, as in the second drawing. Locate points found from the front view on elements or on circular arcs and draw the curve.

CONIC SECTION
(First Drawing)

Problem 1.—Complete the views shown by the intersecting element method. Let $A = 3$, $B = 3\frac{1}{2}$, and $C = 1\frac{1}{8}$ inches.

Problem 2.—Complete the views shown by the horizontal cutting plane method. Assume dimensions for the cone and a suitable location for the plane.

For figures of pyramids see pages 90 and 91.

CONIC SECTION
(Second Drawing)

Problem 1.—Complete the views shown by the horizontal cutting plane method. Let $A = 3$, $B = 3\frac{1}{2}$, $C = 1\frac{1}{8}$ inches, and $\theta = 80°$.

Problem 2.—Complete the views shown by the intersecting element method. Assume dimensions for the cone and a suitable location and angle for the plane.

For figures of pyramids see pages 90 and 91.

PLATE 43

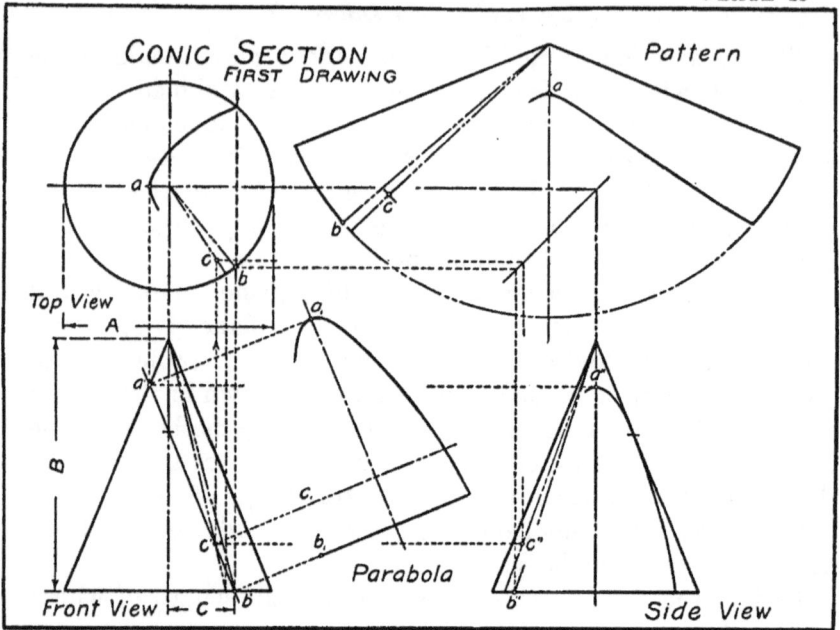

CONIC SECTION
FIRST DRAWING

Pattern

Top View

A

B

Front View ⊢ C ⊣

Parabola

Side View

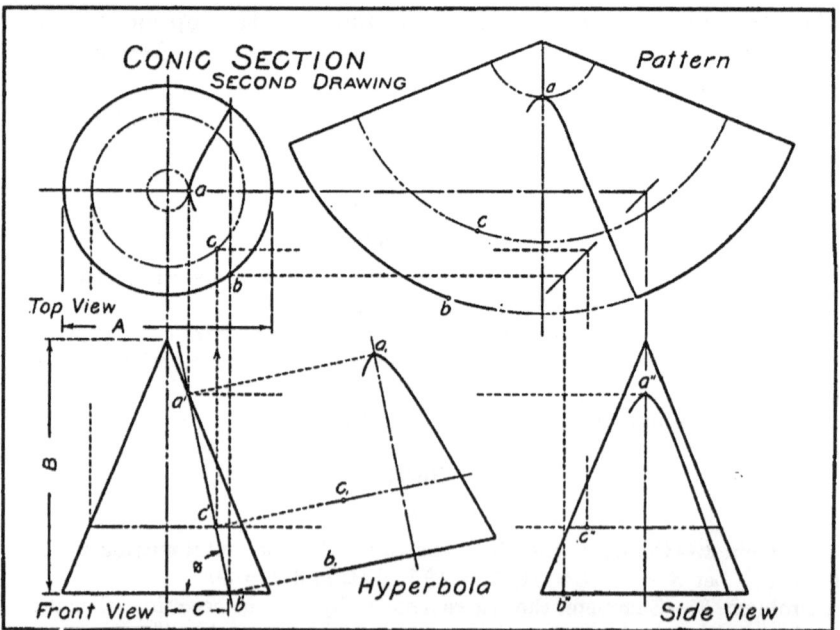

CONIC SECTION
SECOND DRAWING

Pattern

Top View

A

B

Front View ⊢ C ⊣

Hyperbola

Side View

PLATE 44

INTERSECTION

The curve formed by the intersection of two objects is called the line of intersection.

This line may be found by the intersecting elements method, or by the use of cutting planes.   See pages 20 and 21.

INTERSECTING CYLINDERS

The top view of an element passing through point 2_1 in the drawing, will penetrate the vertical cylinder at 2.   This point of penetration projected to the front view of the element will give $2'$, a point on the line of intersection.   Other points are located similarly.

Problem 1.—Find the line of intersection of the first figure when $A = 2\frac{1}{2}$, $B = 3\frac{1}{2}$, $C = 1\frac{3}{4}$, $D = 2$, and $E = 1\frac{1}{2}$ inches.

Find the line of intersection of the second figure when $A = 3$, $B = 3\frac{1}{2}$, $C = 1\frac{3}{4}$, $D = 2\frac{1}{2}$, and $E = 1\frac{1}{2}$ inches.

Problem 2.—Find the line of intersection of the first figure when $A = 3$, $B = 3\frac{1}{2}$, $C = 1\frac{3}{4}$, $D = 3\frac{1}{4}$, and $E = \frac{1}{2}$ inches.   Let the length of the horizontal cylinder be 4 inches.

Find the line of intersection of the second figure when $A = 2\frac{1}{4}$, $B = 3\frac{1}{2}$, $C = 1\frac{3}{4}$, $D = 2\frac{1}{2}$, and $E = \frac{3}{4}$ inches.   Let the length of the horizontal cylinder be $3\frac{3}{4}$ inches.

For additional figure see A, page 92.

INTERSECTING SOLIDS

Points on the line of intersection in the front views are found by assuming the location of the top view of an element and finding its front view.   The point of penetration projected from the top view to the front view will give one point on the line of intersection.

Problem 1.—Find the line of intersection of the first figure when $A = 3$, $B = 3\frac{1}{2}$, $C = 1\frac{3}{4}$, $D = 2\frac{1}{2}$, and $E = 1\frac{1}{2}$ inches.

Find the line of intersection of the second figure when $A = 3$, $B = 3\frac{1}{2}$, $C = 1\frac{3}{4}$, $D = 2\frac{1}{2}$, and $E = 1\frac{1}{2}$ inches.

Problem 2.—Find the line of intersection of the left-hand figure when $A = 2\frac{1}{2}$, $B = 3\frac{1}{2}$, $C = 1\frac{3}{4}$, $D = 2\frac{1}{2}$, and $E = \frac{1}{2}$ inches.   Let the length of the horizontal cylinder be $3\frac{1}{2}$ inches.

Find the line of intersection of the right-hand figure when $A = 2\frac{1}{2}$, $B = 3\frac{1}{2}$, $C = 1\frac{3}{4}$, $D = 2\frac{1}{2}$, and $E = \frac{3}{4}$ inches.   Let the length of the horizontal prism be 4 inches.

PLATE 45

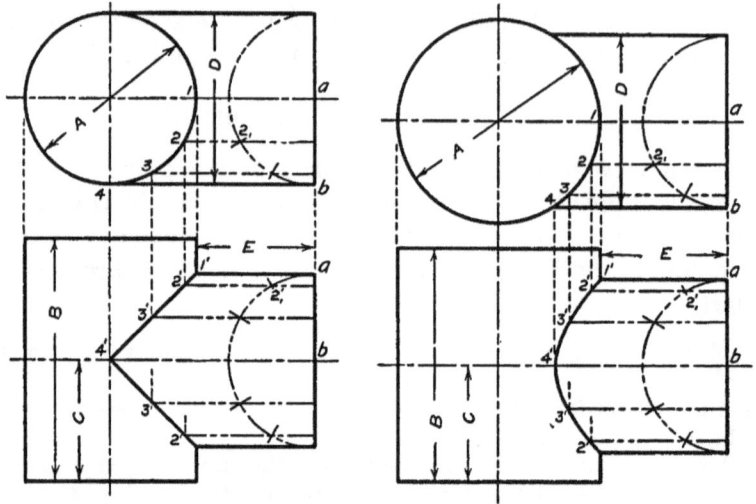

INTERSECTING CYLINDERS

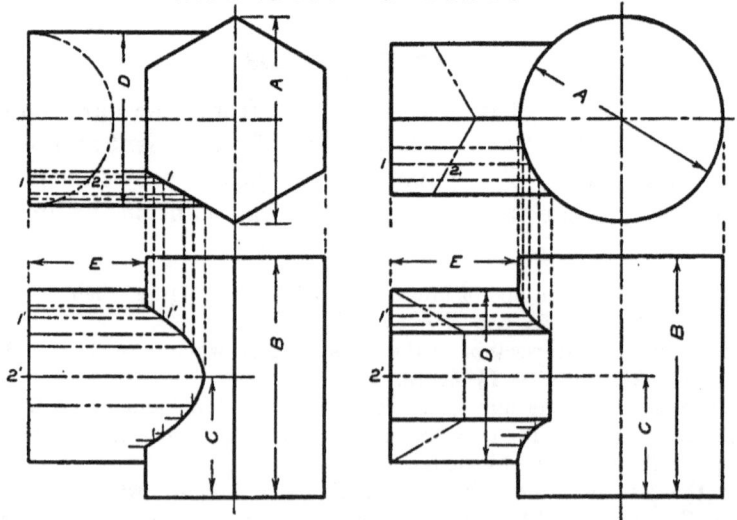

INTERSECTING SOLIDS

PLATE 46

CONE AND CYLINDER

The intersection of the cone and cylinder is found by dividing the base of the cone into any number of equal parts. From these divisions draw the front and the side views of elements of the cone. Project from the point where the elements penetrate the cylinder, shown in the side view, to the corresponding front view projections of these elements. A study of point 2 will make the method clear.

Problem 1.—Find the intersection when $A = 4\frac{1}{2}$, $B = 5$, $C = 2\frac{1}{2}$, $D = 5\frac{1}{2}$, and $E = 1\frac{3}{4}$ inches.

Problem 2.—Find the intersection when $A = 4$, $B = 5$, $C = 2\frac{1}{2}$, $D = 5\frac{1}{2}$, and $E = 1\frac{3}{4}$ inches.

For additional figures see C, page 92; also, A and B, page 95.

CONE AND PRISM

The intersection of the cone and prism is found by assuming a point as 4″ on the end of the prism. Through this point draw the side view of an element of the cone. Find the front view of this element. Project from the point where the element penetrates the prism in the side view, to the front view, giving 4′, a point on the line of intersection. Other points are found similarly.

Problem 1.—Find the line of intersection when $A = 4$, $B = 5$, $C = 2$, $D = 5\frac{1}{2}$, and $E = 2$ inches.

Problem 2.—Find the line of intersection when $A = 4$, $B = 5$, $C = 1\frac{3}{8}$, $D = 5$, and $E = 1\frac{5}{8}$ inches.

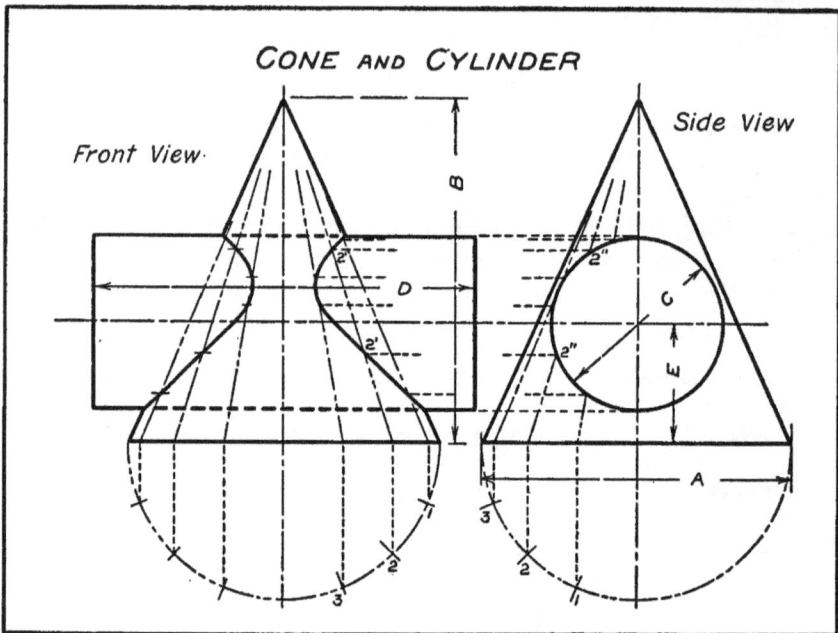

CONE AND CYLINDER

Front View·

Side View

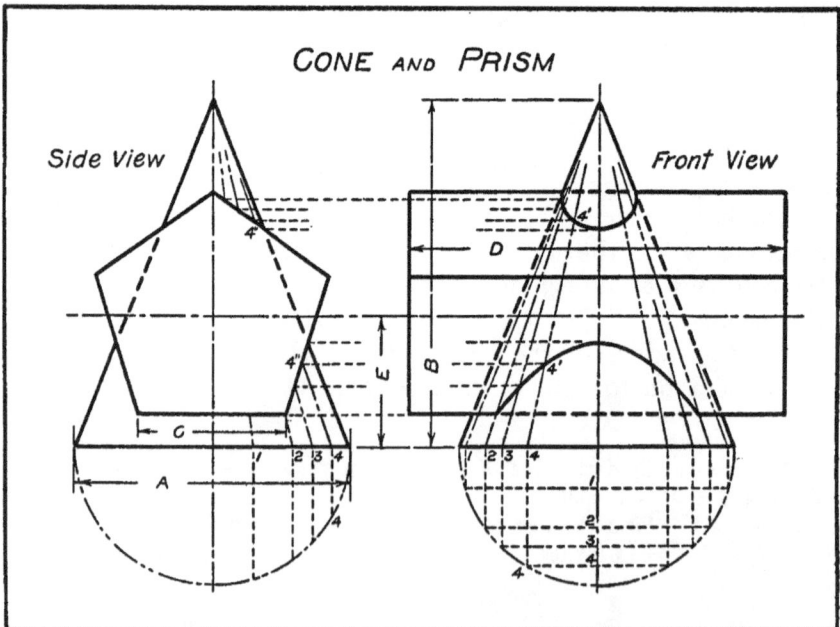

CONE AND PRISM

Side View

Front View

PLATE 48

THREE-PIECE ELBOW

To find the development of a three-piece elbow, first draw the front view and project the top view. Then divide the surface into any number of equal parts, and draw elements. Lay out a surface whose length equals the circumference of the pipe. Divide this length into the same number of parts as the pipe. On vertical lines drawn through the divisions lay off distances equal to the lengths of the elements obtained from the front view, and draw the curves.

The ellipse in the top view is the line of intersection of the horizontal and oblique members of the elbow. It is found by projection from the front view.

Problem 1.—Draw front and top views, and find the development for a three-piece pipe when $A = 2$, $B = 3$, $C = 2\frac{1}{2}$, $D = 1\frac{1}{2}$, and $E = 1$ inches.

Problem 2.—Draw front and top views, and find the development for a three-piece pipe when $A = 2\frac{1}{2}$ inches. Assume suitable dimensions for B, C, D, and E.

For additional figures see A, and B, page 93.

VERTICAL AND OBLIQUE CYLINDERS

To find the intersection of a vertical and an oblique cylinder similar to those shown, first draw the outline of the front view and project the top view. Divide the oblique cylinder, or pipe, into any number of equal parts, and draw elements in both views. One point for the line of intersection is found by projecting from the point where an element pierces the vertical pipe, shown in the top view, to its front view. A study of point 1, in the drawing, will make the method clear.

To develop the half-pattern for the oblique pipe, lay off a surface whose width equals the semi-circumference, and whose length equals the length of the pipe. Divide this surface into the required number of parts. On these parts lay off from mn distances obtained by measurement from the front view.

Problem 1.—Draw front and top views, and find the development for the pipes as shown. Let $A = 2\frac{1}{2}$, $B = 4\frac{1}{2}$, and $C = 1\frac{1}{2}$ inches.

Problem 2.—Draw front and top views, and find the development for two pipes similar to those shown. Let $A = 2$, $B = 4\frac{1}{2}$, and $C = 1\frac{3}{4}$ inches.

For additional figures see B and D, page 92.

PLATE 49

THREE-PIECE ELBOW

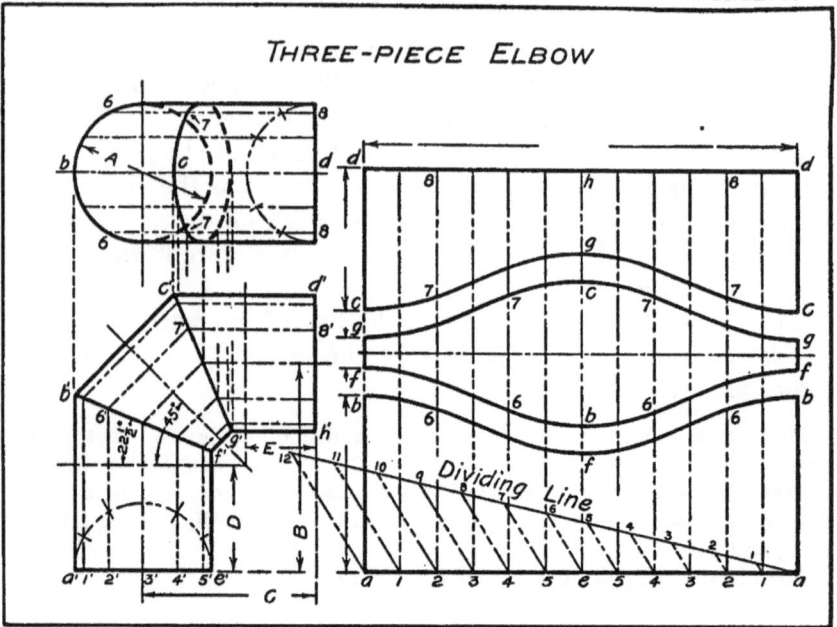

Dividing Line

VERTICAL AND OBLIQUE CYLINDERS

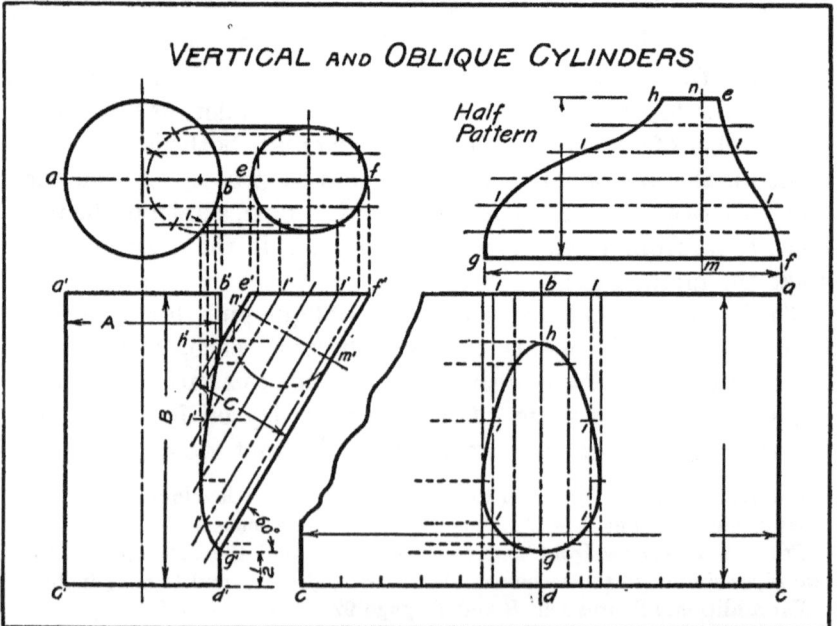

Half Pattern

PLATE 50

OFFSET PIPES AND ELBOWS

Offset pipes and elbows may be made from straight or tapering pipes by cutting at suitable angles as shown in the drawings.

To make an offset pipe, select two points on the axis of the pipe, thus dividing it into three parts. Lay out these parts of the axis to give the amount of offset wanted. Measure the angle the inclined axis makes with the original axis. This angle divided by two will give the angle of cut.

The angle of cut for a three-piece elbow will be one-half the angle the oblique section makes with the uncut pipe.

CIRCULAR OFFSET PIPE

Problem 1.—Draw an offset pipe, and find the development when $A = 1\frac{3}{4}$, $B = 1\frac{1}{4}$, $C = 4$, $D = 1\frac{1}{2}$, and $E = 2\frac{1}{4}$ inches.

Problem 2.—Draw an offset pipe, and find the development when $A = 1\frac{1}{2}$ and $C = 4$ inches. Let $\theta = 45°$. Assume the other dimensions.

For additional figures see A, B, and C, page 93.

THREE-PIECE CONICAL ELBOW

Problem 1.—Draw a three-piece right-angled conical elbow, and find the development when $A = 2\frac{3}{4}$, $B = 1\frac{1}{8}$, $C = 2\frac{1}{2}$, $D = 3$, and $E = 1\frac{1}{4}$ inches.

Problem 2.—Draw a conical offset pipe, and find the development when $A = 2\frac{1}{2}$ and $B = 1\frac{1}{4}$ inches. Let $\theta = 60°$. Assume the other dimensions and the amount of offset.

For additional figures see D, E, and F, page 93.

PLATE 51

CIRCULAR OFF-SET PIPE

Patterns

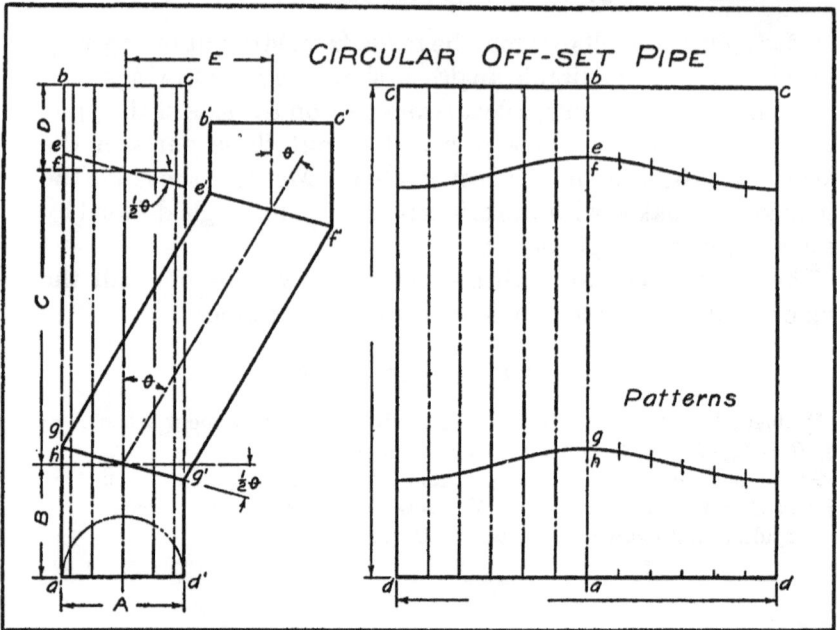

THREE-PIECE CONICAL ELBOW

Patterns

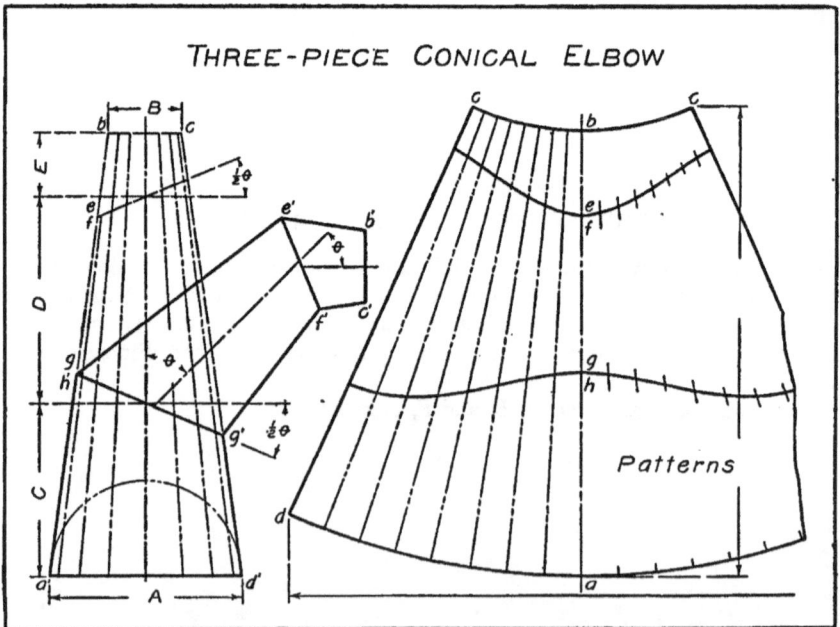

PLATE 52

INTERSECTING PIPES

The intersection of two pipes of equal diameters with intersecting axes, oblique or at right angles to each other, is shown by straight lines in the front view.   .

The intersection of two pipes of unequal diameters with axes intersecting or offset, oblique or at right angles, is shown by curved lines in the front view.

The ellipses in the top views of the oblique pipes may be found by projection, as in previous problems, or may be drawn after the lengths of the axes have been determined, as explained for Figs. 31 or 32.

A study of the reference letters and figures in the drawings should enable the student to work the following problems:

INTERSECTING PIPES
(Same Diameters)

Problem 1.—Find the development for two intersecting pipes of equal diameters and intersecting axes. Let $\theta = 45°$, $A = 2\frac{1}{4}$, $B = 4$, $C = \frac{1}{2}$, and $D = 3$ inches.

Problem 2.—Find the development for two intersecting pipes of equal diameters and intersecting axes. Let $\theta = 60°$ and $A = 2$ inches. Assume other suitable dimensions.

For additional figures see *A* and *B*, page 92.

INTERSECTING PIPES
(Different Diameters)

Problem 1.—Find the development for two intersecting pipes of unequal diameters and non-intersecting axes. Let $\theta = 45°$, $A = 2\frac{1}{2}$, $B = 4$, $C = 1\frac{1}{4}$, $D = 4\frac{3}{4}$, $E = \frac{3}{4}$, and $F = \frac{3}{8}$ inches.

Problem 2.—Find the development for two intersecting pipes of unequal diameters and non-intersecting axes. Let $\theta = 45°$, $A = 2\frac{1}{4}$, $C = 1\frac{1}{2}$, and $F = \frac{1}{4}$ inches. Assume other suitable dimensions.

For additional figures see *C* and *D*, page 92.

PLATE 53

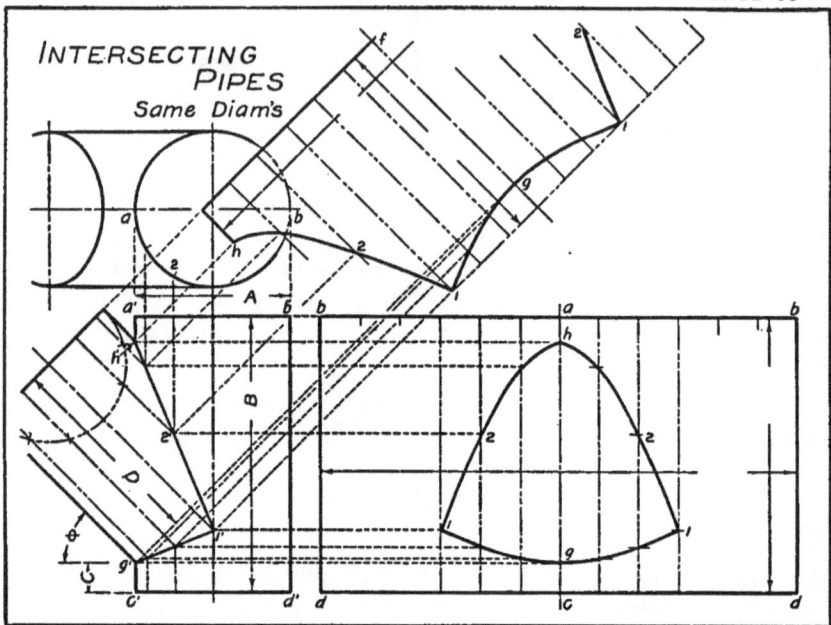

PLATE 54

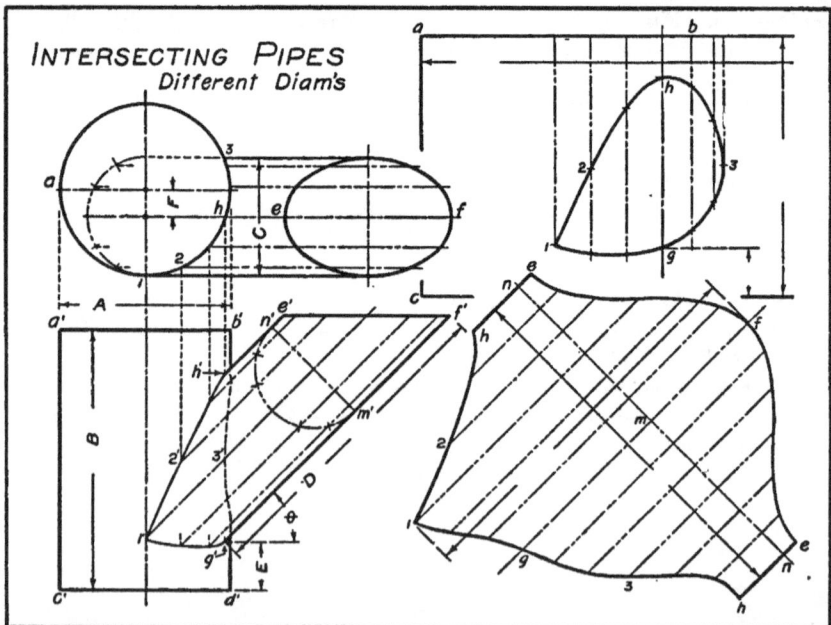

TRANSITION PIECE
Octagonal Outlet

To find the development of transition piece similar to that shown, draw top and front views. Find true dimensions of the surfaces, of which there are two kinds, and lay out the pattern as shown.

The surface *adeb* in the development is shown in its true height by f_1b' in the front view, while the lengths *de* and *ab* are found from the top view.

Problem 1.—Draw a transition piece having a square inlet and an octagonal outlet, and find the development when $A = 2\frac{1}{2}$ and $B = 3$ inches. Let the outlet be as shown.

Problem 2.—Draw a transition piece having a square inlet and a $1\frac{3}{4}$-inch square outlet and find the development when $A = 2\frac{3}{4}$ and $B = 3\frac{1}{4}$ inches.

TRANSITION PIECE
Round Outlet

To find the development of a transition piece similar to that shown, draw top and front views. Divide *ed* into four equal parts and project to the front view. Construct a true lengths diagram, as follows: On a horizontal line lay off lengths *be*, *b*1, and *b*2 and erect vertical lines of lengths equal to the height of the object. The true lengths are then drawn as shown. Draw the triangle *abe*. From *e* draw an arc of radius *e*1, found in the top view, and from *b* draw an intersecting arc of radius $b1_1$, found from the diagram. The intersection of these arcs will be one point on the required curve. Other points are found similarly. This is called the *triangulation* method of development.

Problem 1.—Find the development for a transition piece having a square inlet and a round outlet when $A = 2\frac{1}{2}$, $B = 3$, and $C = 1\frac{3}{4}$ inches.

Problem 2.—Find the development for a transition piece having a square inlet and a round outlet when $A = 2\frac{1}{2}$, $B = 3$, and $C = 3$ inches.

PLATE 55

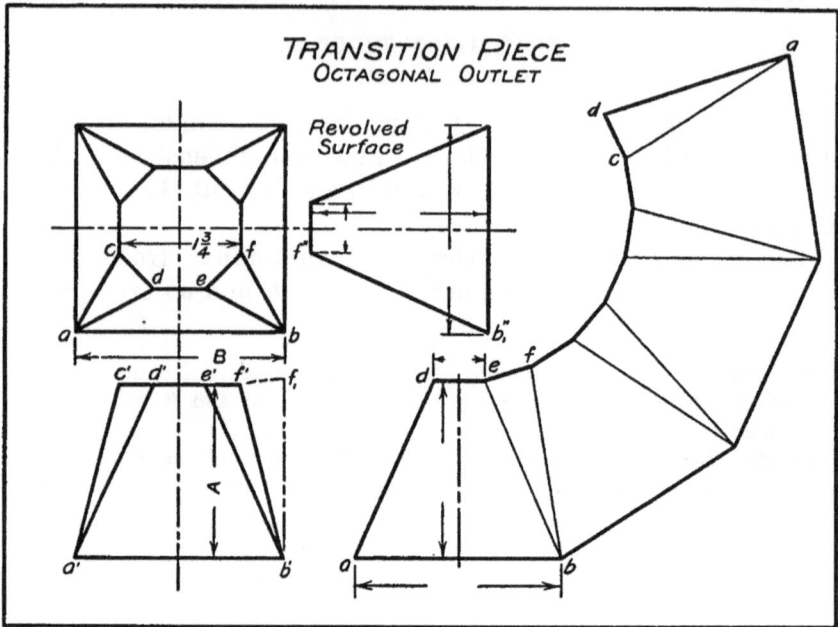

TRANSITION PIECE
OCTAGONAL OUTLET

Revolved
Surface

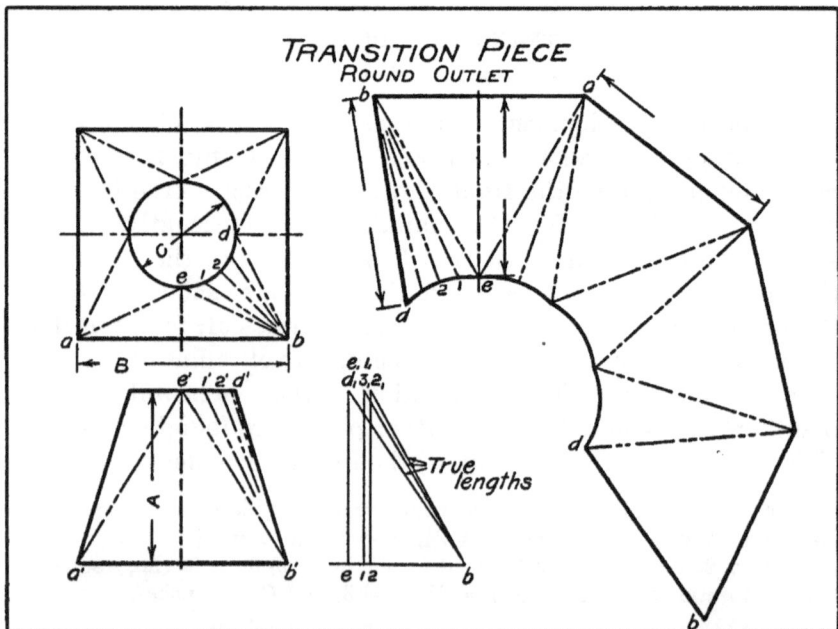

TRANSITION PIECE
ROUND OUTLET

True
lengths

PLATE 56

SCALENE CONE

To find the development of a scalene cone, draw front and top views. Divide the base into any number of equal parts. Lay out a true lengths diagram and find the development by the intersecting arc method. A study of the drawing will make the solution clear.

Problem 1.—Find the development for a scalene cone with a circular base when $A = 3$, $B = 3\frac{1}{4}$, and $C = 2\frac{3}{4}$ inches.

Problem 2.—Find the development for the frustum of a scalene cone when $A = 2\frac{3}{4}$, $B = 3\frac{1}{2}$, $C = 3$, and $D = 1\frac{3}{4}$ inches.

TRANSITION PIECE

To find the development of the transition piece shown, draw top and front views. Divide one half of the top view into an even number of parts, project to the front view, and draw elements and diagonals as shown. Lay out the element and the diagonal diagrams by laying off on horizontal lines the lengths measured from the top view, as bases of right-angled triangles, whose heights equal the height of the object. The true lengths of the elements and the diagonals will be equal to the hypotenuses of these triangles.

Find the development by beginning at b, an assumed point, and locating points 10, 9, 8, etc., by the intersecting arc method. To locate point 10, for instance, draw an arc from b as center and of radius $b10$, found from the top view; also, draw an intersecting arc from d as center and of radius $o10$, found from the diagram of diagonals. The intersection of the arcs will give point 10. Other points are found similarly.

Problem 1.—Find the development for the transition piece when $A = 4$, $B = 3$, $C = 2\frac{1}{2}$, $D = 1\frac{3}{4}$, and $E = 2$ inches.

Problem 2.—Find the development for the transition piece when $A = 4$, $B = 3\frac{1}{2}$, $C = 2\frac{1}{2}$, $D = 2$, and $E = 1\frac{3}{4}$ inches.

PLATE 57

SCALENE CONE

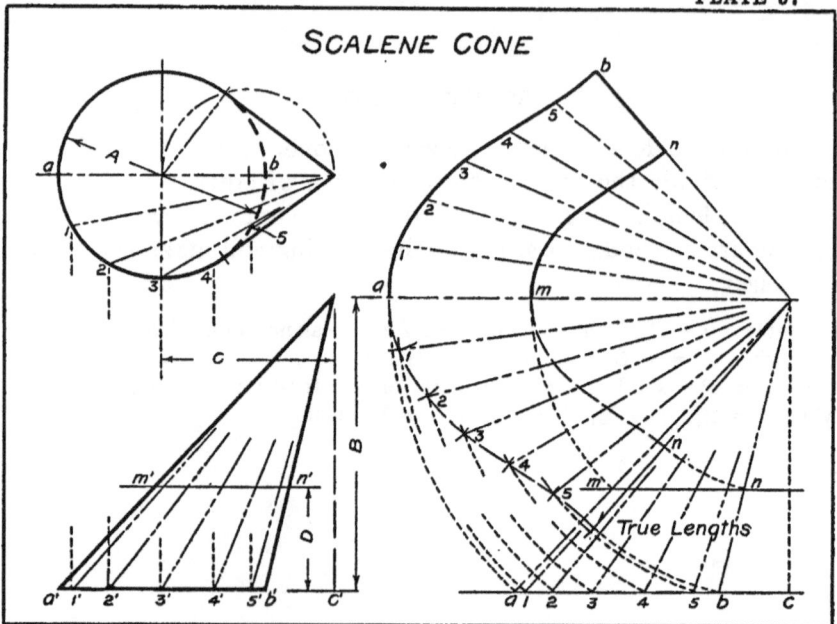

True Lengths

TRANSITION PIECE·

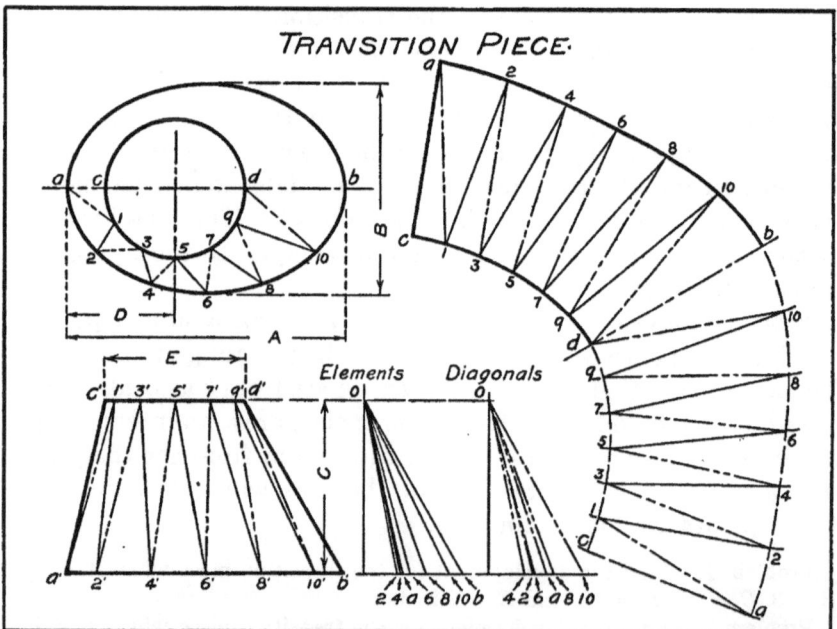

Elements   Diagonals

PLATE 58

INTERSECTION OF TWO PRISMS

To find the projections of two intersecting prisms as shown, proceed in the following order: Draw top and front views of B; the axis of A; the auxiliary view of B, on axis at right angles to axis for A; the auxiliary view of A; the front and the top views of A. Letter the ends of the edges as shown.

To find one point on the intersection, proceed as follows: Select a point, as 1, in the top view. This point will be 1_1 in the auxiliary view. The intersection of projections from 1 and 1_1 will give $1'$, a point on the intersection. Projections from 2 and 2_1 on edge a, will give $2'$, another point on the intersection. A straight line drawn from $1'$ to $2'$ will be a portion of the intersection. Then find $3'$ and join it to $2'$ by a straight line. Next find $4'$ and continue this operation until the complete line of intersection is found.

Problem 1.—Find the line of intersection of the two prisms as shown. Ascertain the dimensions and angles by measuring the drawing. The scale of the drawing is ⅜ inch equals 1 inch.

Problem 2.—Find the line of intersection of the two prisms when $\alpha = 10°$, $\theta = 20°$, and $\phi = 30°$. Ascertain the dimensions of the prisms and location of the oblique axis by measuring the drawing. The scale of the drawing is ⅜ inch equals 1 inch.

For additional figures see page 94.

DEVELOPMENT OF TWO INTERSECTING PRISMS

Find the development of prism A, in the upper drawing, by laying out a surface equal to the surface of the prism. Letter the surface corresponding to the edges. Point 1, on edge i, is found by measuring its distance from i in the front view and laying this off from i in the development. Point 2 is found by measuring the distance of 2_1 from i_1 in the auxiliary view, and the distance from line $j'h'$ in the front view, and laying off these distances in the development. Other points are found similarly.

The development for prism B is found by a similar method.

Problem 1.—Find the development of the prisms for Problem 1 in the INTERSECTION OF TWO PRISMS.

Problem 2.—Find the development of the prisms for Problem 2 in the INTERSECTION OF TWO PRISMS.

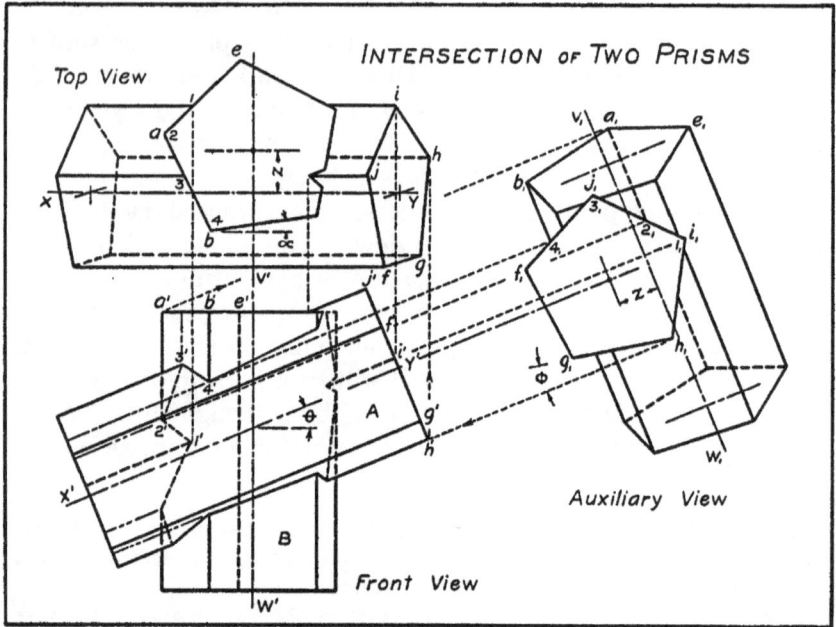

INTERSECTION OF TWO PRISMS

Top View

Auxiliary View

Front View

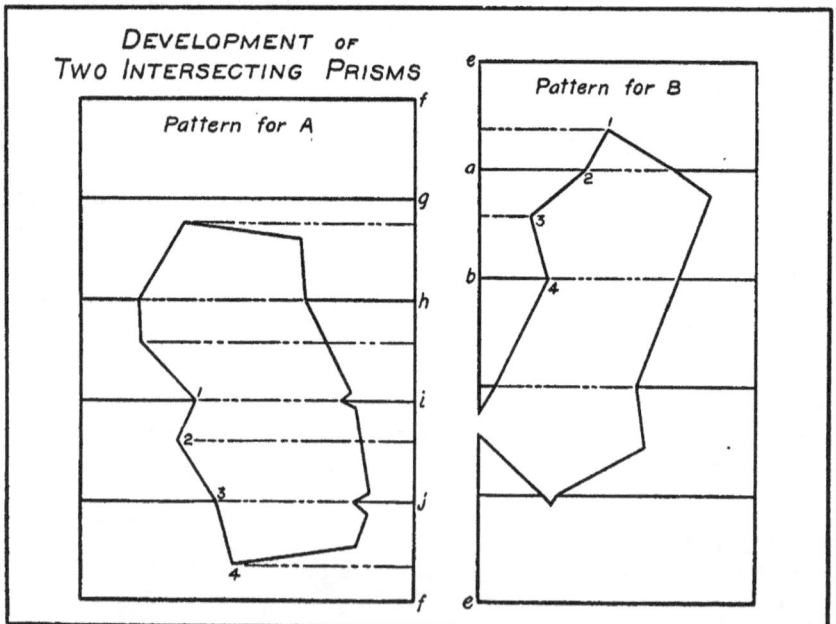

DEVELOPMENT OF
TWO INTERSECTING PRISMS

Pattern for A

Pattern for B

PLATE 60

INTERSECTION OF TWO CONES
First Method

Cutting Plane Method.—The intersection of two cones may be found by means of a number of cutting planes, each containing elements of both surfaces. The intersection of elements in the same plane will give points on the line of intersection.

Draw a line of indefinite length through the apexes *a* and *e*. Draw lines through the bases, locating points *k*, *x*, and *y*. From *k*, draw lines at right angles to the bases. From *d* and *h*, as centers, draw half-circles as shown.

Select any point, as *q*, on the revolved view of the base of the smaller cone, and draw a line from *x* through *q*, giving points *p* and *n*. With *k* as center, draw an arc from *n* to *n'* and a line from *n'* to *y*, giving points *r* and *s* on the revolved view of the base of the larger cone.

The lines drawn from *x* and *y* to *n* and *n'* are the traces of a plane passing through the apexes of the cones and will cut elements from both cones. The intersections of these elements will locate four points on the line of intersection. Two of these are shown by points 1 and 2. Other points on the intersection may be found by drawing additional planes from *o*, *m*, and *l*.

A study of the drawing should enable the student to find the intersection in the end view.

Problem 1.—Find the line of intersection in the front and the side views of the cones as shown. For dimensions, measure the drawing to a scale of $\frac{3}{8}$ inch equals 1 inch.

Problem 2.—Find the line of intersection in the front and the side views of two cones, similar to those shown, the axes intersecting and at right angles to each other. For dimensions, measure the drawing to a scale of $\frac{3}{8}$ inch equals 1 inch.

For additional figures see *C*, *D*, *E*, and *F*, page 95.

DEVELOPMENT OF TWO INTERSECTING CONES

Draw sectors equal to the convex surfaces of cones *A* and *B*. Lay off points *r*, *s*, and *p*, obtained from the bases of the front view, and draw the elements. On these elements locate points 1 and 2, their true distance from *r*, *s*, and *p* being found from the front view. Additional points are found similarly.

Problem 1.—Find the developments for Problem 1, INTERSECTION OF TWO CONES, First Method.

Problem 2.—Find the developments for Problem 2, INTERSECTION OF TWO CONES, First Method.

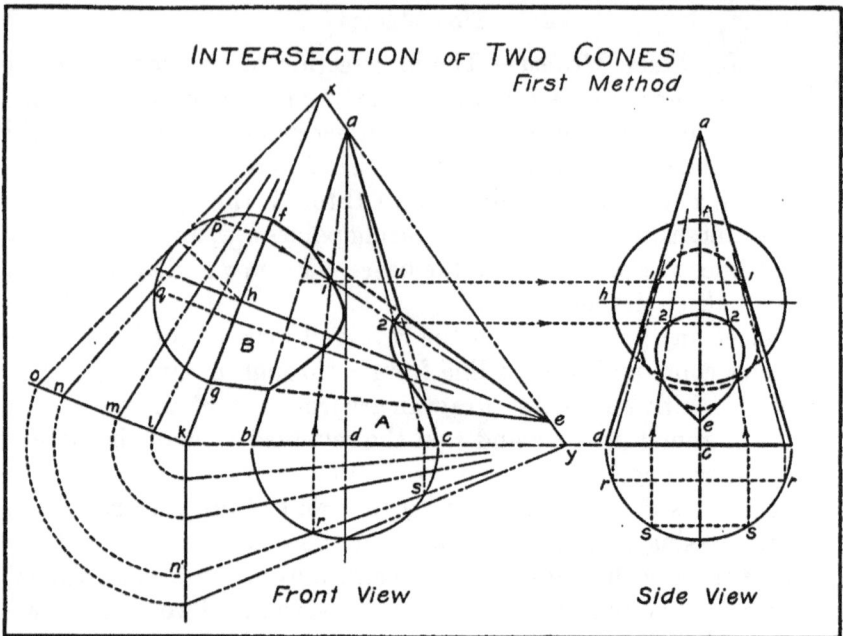

INTERSECTION OF TWO CONES
First Method

Front View   Side View

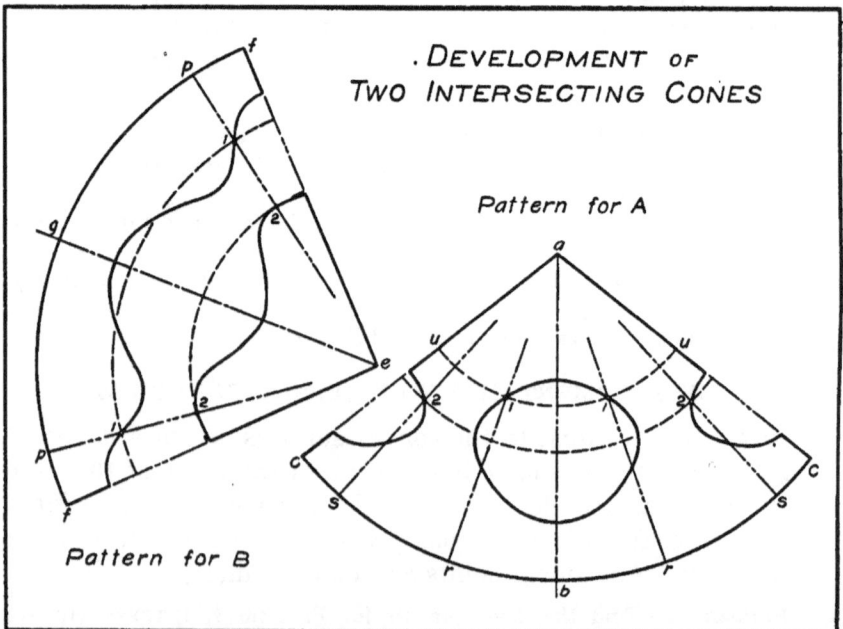

DEVELOPMENT OF
TWO INTERSECTING CONES

Pattern for A

Pattern for B

PLATE 62

INTERSECTION OF TWO CONES
Second Method

Concentric Sphere Method.—The projections of concentric spheres, drawn with the intersection of the axes of two cones as a center, will cut circles from both surfaces, where the spheres and cones intersect. The intersections of these circles will give points on the line of intersection.

With *o* as center, draw the projection of a sphere of an assumed radius *op*, cutting cone *A* at *r* and *t*, and cone *B* at *q* and *s*. Lines drawn from these points, at right angles to the axes of the cones, show the vertical projections of four circles, two of which are parallel to the base of *A*, and two parallel to the base of *B*. The intersections of these circles give points 1, 2, and 3 on the line of intersection. Point 4 is found by drawing a portion of a sphere of an assumed radius *oy*.

A study of the drawing should enable the student to find additional points in the front view, also the points in the top and the side views.

Problem 1.—Find the line of intersection of the cones as shown. For dimensions, measure the drawing to a scale of ⅜ inch equals 1 inch.

Problem 2.—Find the line of intersection in three views of the two cones when the axes are at right angles to each other. For dimensions, measure the drawing to a scale of ⅜ inch equals 1 inch.

For additional figures see *C*, *D*, *E*, and *F*, page 95.

DEVELOPMENT OF TWO INTERSECTING CONES

Draw a sector equal to the convex surface of cone *B*. Draw the elements finding their location from the revolved view of the base of the cone as shown in the front view. Circular arcs with radii measured on *kf* in the front view, and drawn in the pattern, will intersect the elements giving points on the required curve as shown.

To locate the elements for pattern *A*, it will be necessary to draw a revolved view of the base, similar to that for *B*, to find the location of elements drawn through the points on the line of intersection.

Problem 1.—Find the developments for Problem 1, INTERSECTION OF TWO CONES, Second Method.

Problem 2.—Find the developments for Problem 2, INTERSECTION OF TWO CONES, Second Method.

PLATE 63

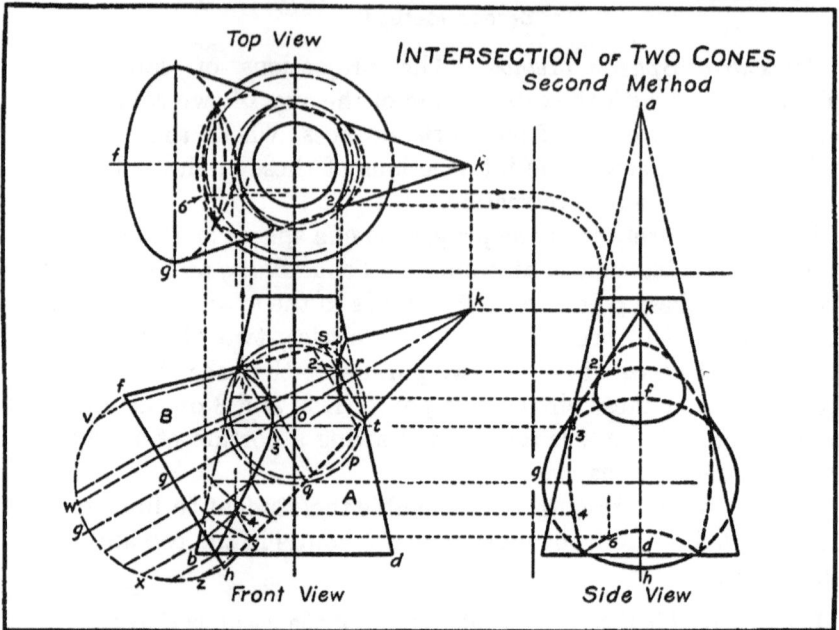

Top View

INTERSECTION of TWO CONES
Second Method

Front View

Side View

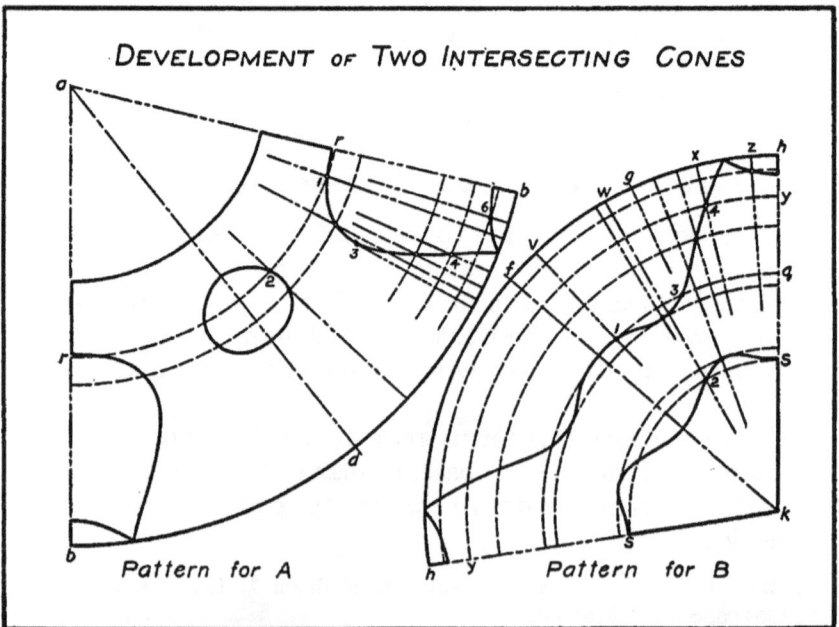

DEVELOPMENT of TWO INTERSECTING CONES

Pattern for A

Pattern for B

PLATE 64

INTERSECTION OF TWO CONES
Third Method

Revolved Section Method.—Points on the intersection of two cones may be found by the intersection of lines shown by a number of revolved sections.

A revolved section, produced by a plane cutting a triangle from the one and a curved line from the other of two intersecting cones, will show two, or four, points on the line of intersection.

Divide the revolved view of *gh* into eight equal parts and project these points of division to *gh*, giving points *p*, *l*, and *n*. From *f* draw plane traces through *p*, *l*, and *n*, giving points *r*, *s*, and *t*. Divide the bottom view of cone *A* into equal parts and draw elements as shown. From the intersections of *fs* and the elements cut, also from *l*, draw lines at right angles to *fs*. Draw *f's'* parallel to *fs*. On *f's'* as an axis, find the revolved section cut from *A* by the plane *fs*. From *f'*, through *k''l''*, equal to *k'l'*, draw lines intersecting the revolved section of *A* at points 2-2. These points projected back to *fs* will be two points on the line of intersection. Additional points are found from the revolved sections *C* and *E*.

Problem 1.—Find the line of intersection of the cones as shown. For dimensions, measure the drawing to a scale of ⅜ inch equals 1 inch. Let the axes make an angle of 30° with each other.

Problem 2.—Find the line of intersection of two cones, similar to those shown, whose axes make an angle of 45° with each other. For dimensions, measure the drawing to a scale of ⅜ inch equals 1 inch.

For additional figures see *C*, *D*, *E*, and *F*, page 95.

DEVELOPMENT OF TWO INTERSECTING CONES

Draw the front view showing the line of intersection, and the elements *v*, *w*, and *x*. These elements are found by drawing lines through the points found on the line of intersection in the upper drawing.

A study of the drawings should enable the student to complete the developments.

Problem 1.—Find the developments for Problem 1, INTERSECTION OF TWO CONES, Third Method.

Problem 2.—Find the developments for Problem 2, INTERSECTION OF TWO CONES, Third Method.

PLATE 65

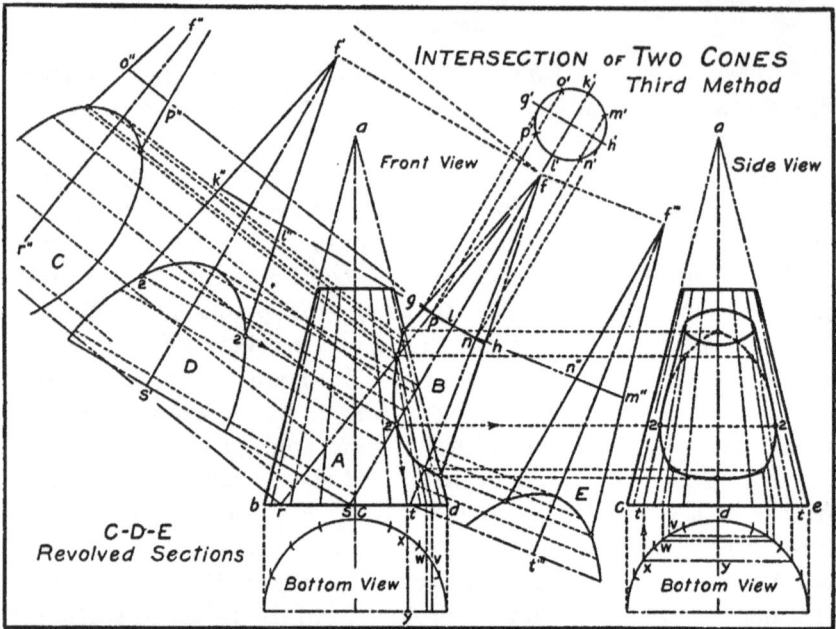

INTERSECTION of TWO CONES
Third Method

Front View

Side View

C-D-E
Revolved Sections

Bottom View

Bottom View

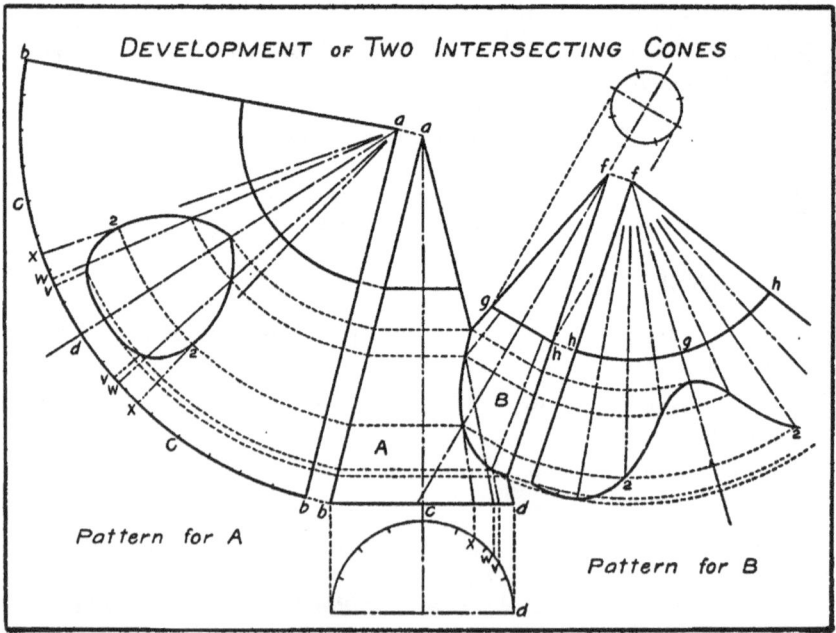

DEVELOPMENT of TWO INTERSECTING CONES

Pattern for A

Pattern for B

PLATE 66

TRIANGULAR PYRAMID

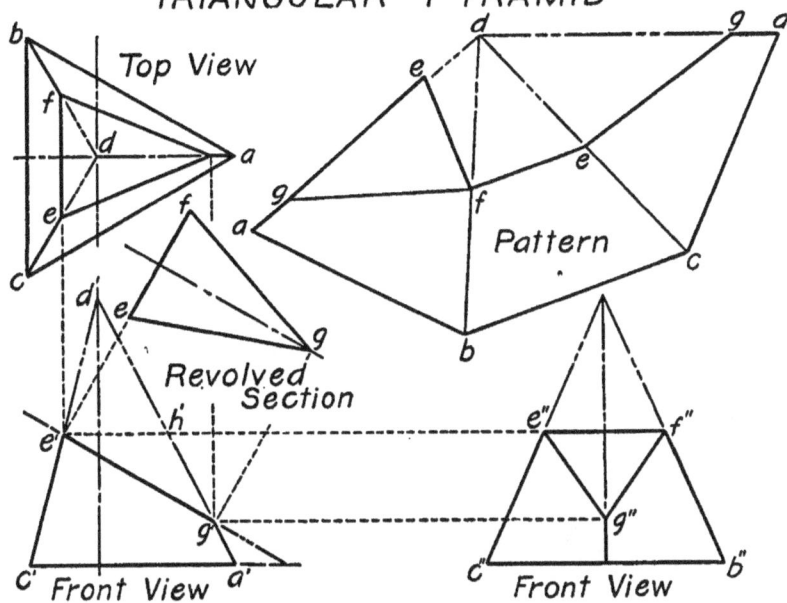

Top View

Pattern

Revolved Section

Front View

Front View

PENTAGONAL PYRAMID

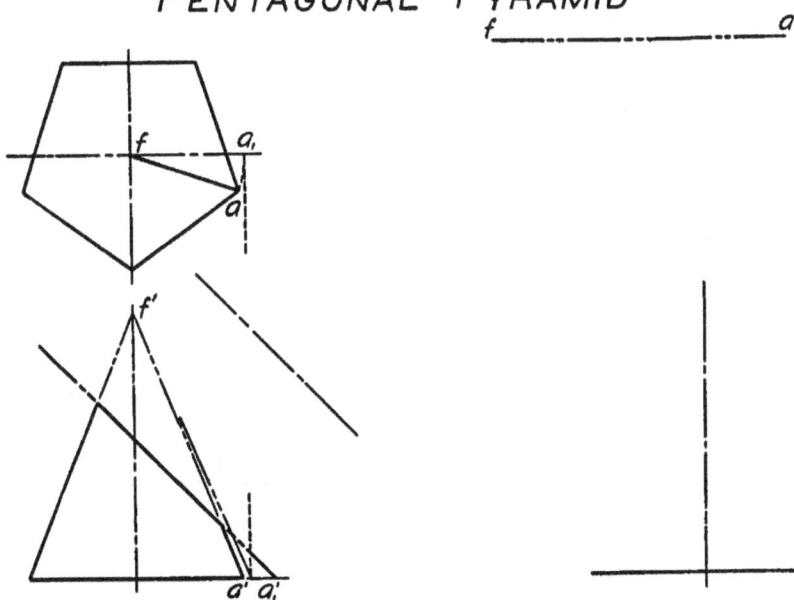

Supplementary problems.

SQUARE PYRAMID

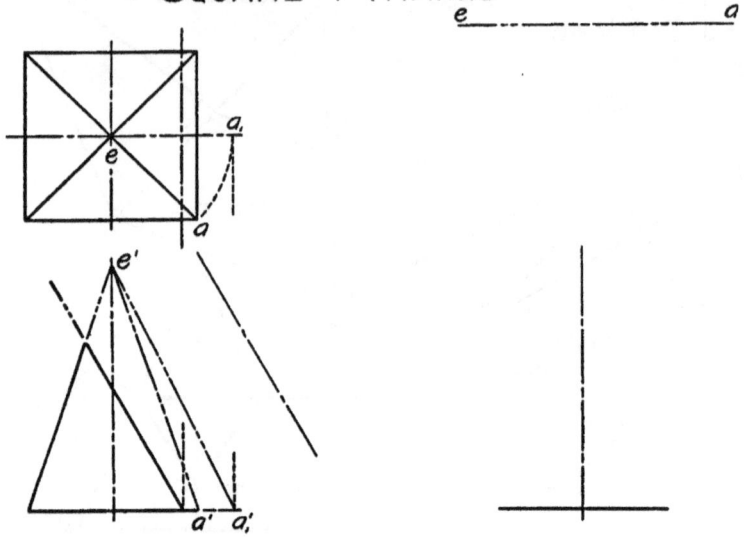

HEXAGONAL PYRAMID

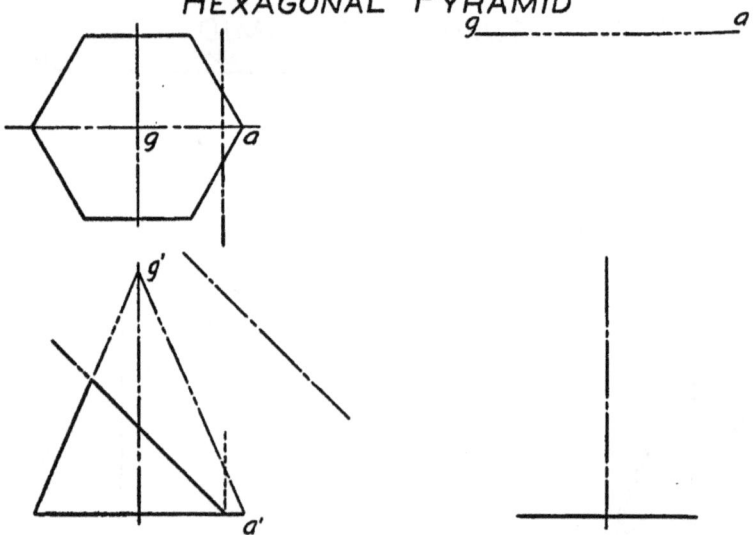

Supplementary problems

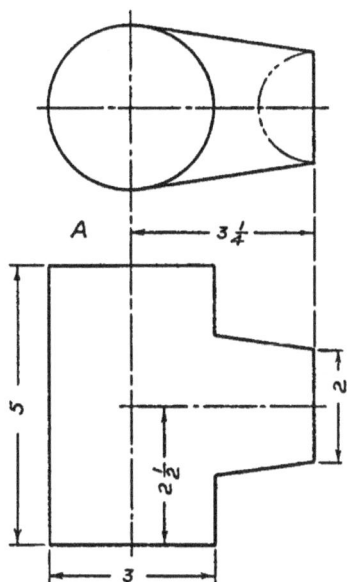

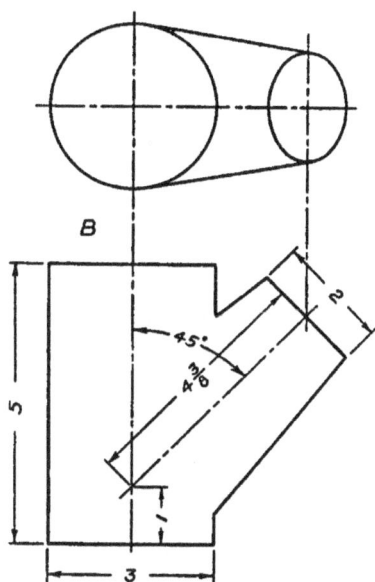

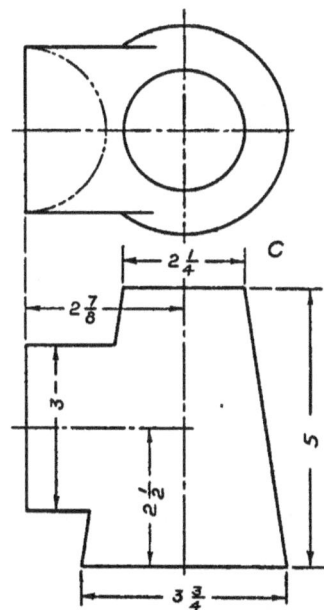

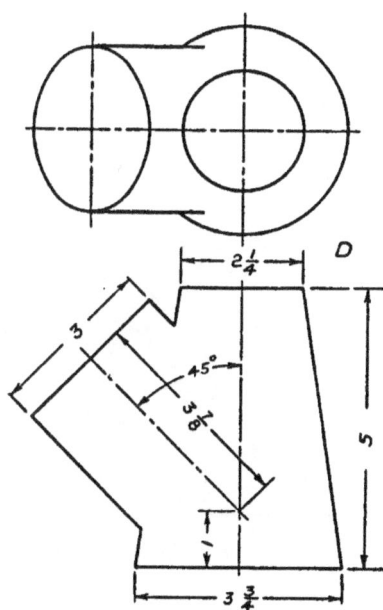

Supplementary problems.

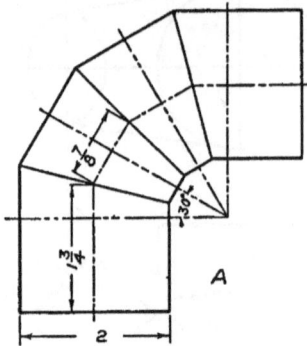

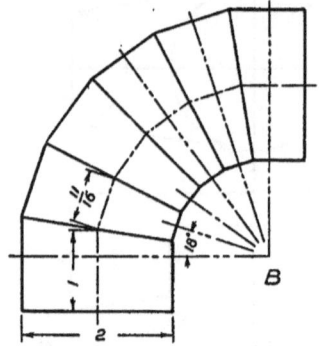

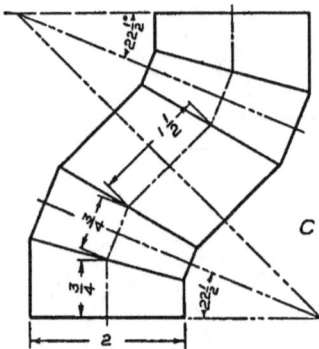

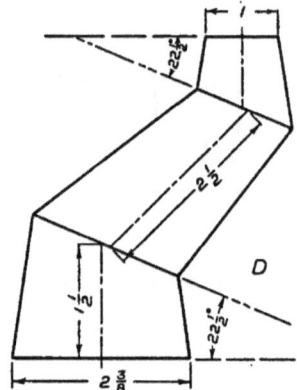

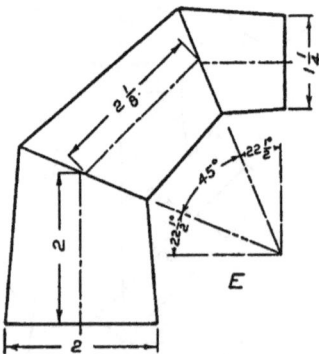

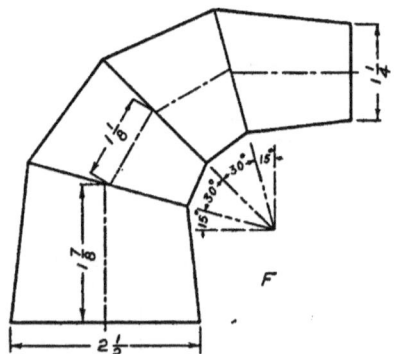

Supplementary problems.

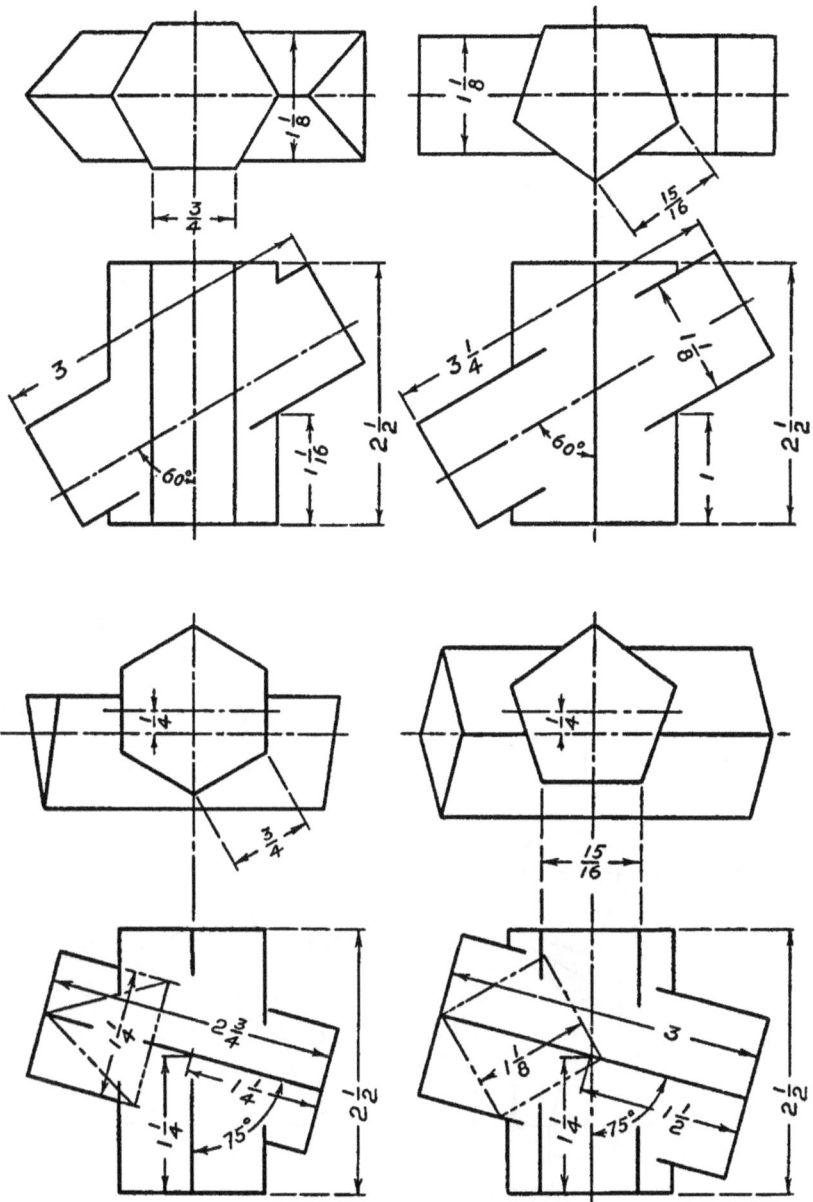

Supplementary problems.

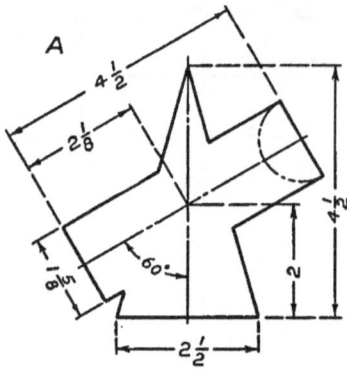

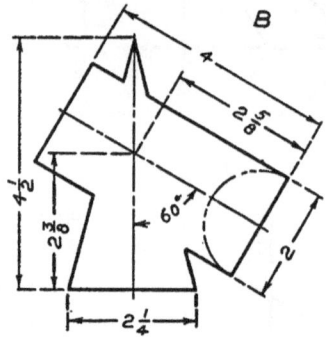

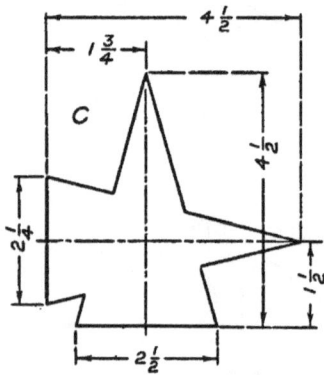

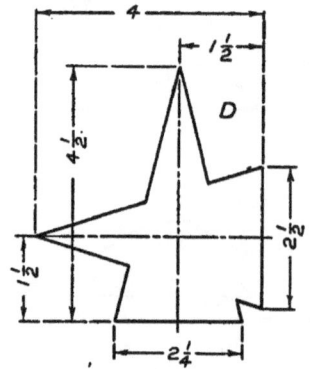

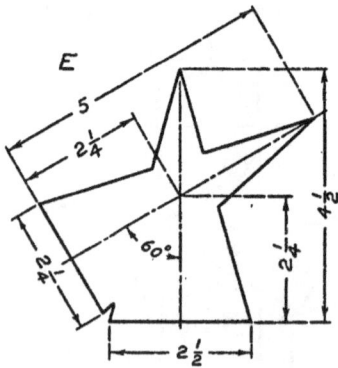

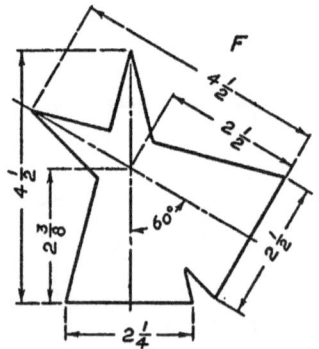

Supplementary problems.

SECTION III

ISOMETRIC AND OBLIQUE DRAWING

ISOMETRIC DRAWING

Isometric drawing is a branch of mechanical drawing which shows three faces of an object in one view, and is based on the following Principles:

1. Three axes are drawn from a common point, called the origin, on a horizontal line; one is drawn vertical, one 30° to the left, and one 30° to the right.

2. The axes represent lines which are mutually perpendicular to each other, and on them length, width, and height may be measured.

3. Measurements of length, width, and height can be made only on the axes, or on lines parallel to the axes.

4. Lines which are parallel in an object are parallel in the drawing.

5. Lines which are vertical in an object are vertical in the drawing, and will be drawn in their true lengths.

6. Lines which are not parallel to one of the isometric axes are said to be non-isometric, and cannot be measured in their true lengths.

7. Non-isometric lines are located by offsets, and may be either longer or shorter than their true measurements.

8. Lines which are at right angles in an object are shown at 60° or 120° to each other in the drawing.

9. Isometric angles cannot be measured in degrees.

10. An isometric circle is an ellipse, and a quarter-circle will be shown in an angle of either 60° or 120°.

ISOMETRIC BLOCKS

Problem 1.—Make a drawing of the figures shown beginning with *a* at the origin of the axes.

Problem 2.—Make a drawing of the figures shown beginning with *b* at the origin of the axes.

JOINTS

Problem 1.—Make a drawing showing the joints beginning with *a* at the origin of the axes.

Problem 2.—Make a drawing showing the joints beginning with *b* at the origin of the axes.

PLATE 67

ISOMETRIC BLOCKS

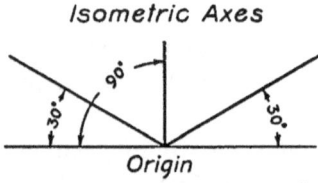

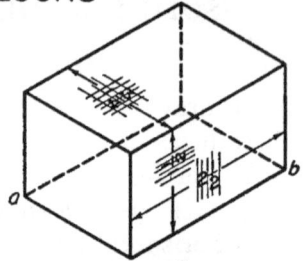

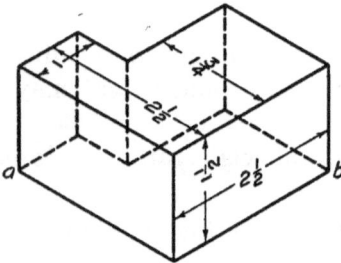

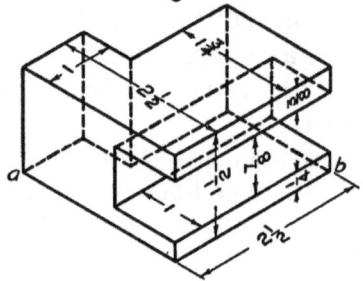

Isometric Axes

Origin

JOINTS

MORTISE AND TENON

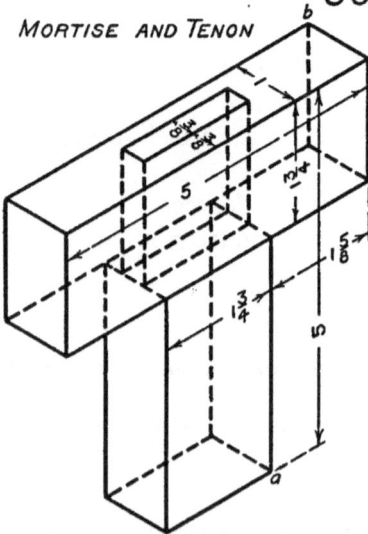

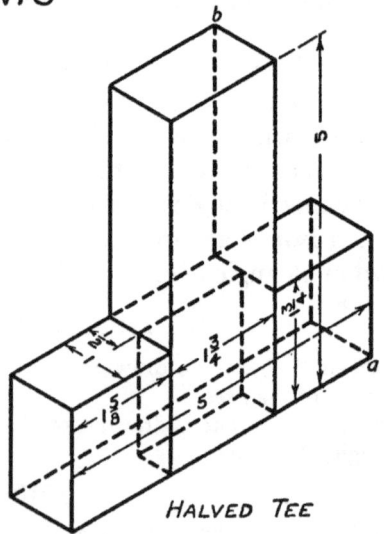

HALVED TEE

PLATE 68

JOINTS

This drawing shows two joints in which the vertical member of the left-hand figure is withdrawn vertically through a distance of $\frac{7}{8}$ inch, while the vertical member of the right-hand figure is withdrawn horizontally through a distance of $1\frac{1}{8}$ inches.

Problem 1.—Make a drawing of the figures shown beginning with the corners *a* at the origin of the axes.

Problem 2.—Make a drawing of the figures shown beginning with the corners *b* at the origin of the axes.

MORTISE AND TENON JOINTS

This drawing shows two mortise and tenon joints in which the horizontal member of the left-hand figure is withdrawn 1 inch, while the vertical member of the right-hand figure is withdrawn $1\frac{1}{2}$ inches.

Problem 1.—Make a drawing showing the joints beginning with the corners *a* at the origin of the axes.

Problem 2.—Make a drawing showing the joints beginning with the corners *b* at the origin of the axes.

PLATE 69

JOINTS

HALVED DOVETAIL

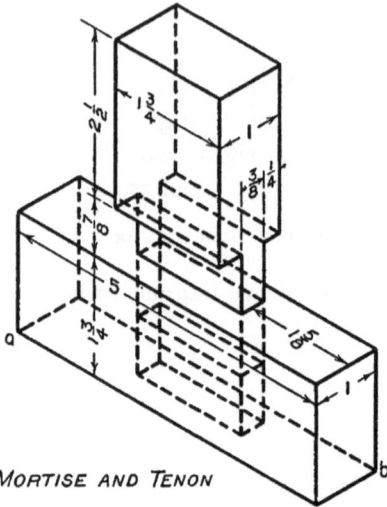

MORTISE AND TENON

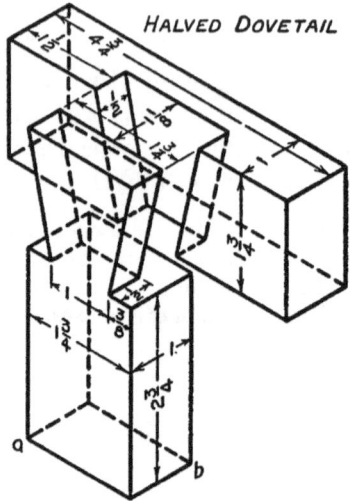

MORTISE AND TENON JOINTS

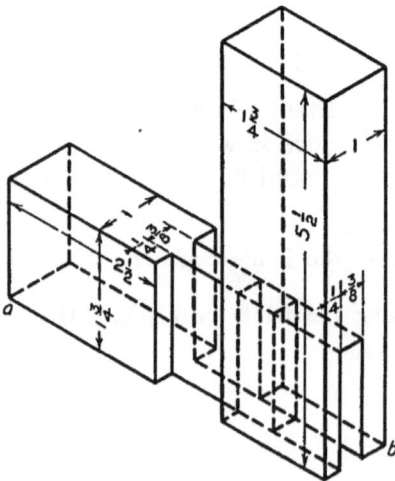

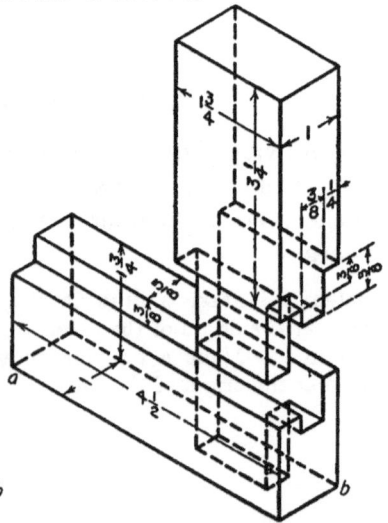

PLATE_70

MITER BOX

This drawing shows the top view, and an isometric drawing of a common miter box.

The oblique lines, showing the saw cuts in the isometric drawing, are located by laying off, on both edges of the upper surface, the measurements found from the top view.

Problem 1.—Draw the top view of the object as shown, and make an isometric drawing with the corner *a* at the origin of the axes.

Problem 2.—Draw the top view of the object as shown, and make an isometric drawing with the corner *b* at the origin of the axes.

DRAWER AND TABLE JOINTS

This drawing shows a unmber of joints such as may be used in drawer and table constructions.

Problem 1.—Make a drawing of the joints, showing *A*, *B*, and *C* assembled, omitting hidden lines.   Give all dimensions.

Problem 2.—Make a drawing of the joints, showing *D* and *E* assembled, omitting hidden lines.   Give all external dimensions.

PLATE 71

MITER BOX
Scale $\frac{3}{4}$ in. = 1 in.

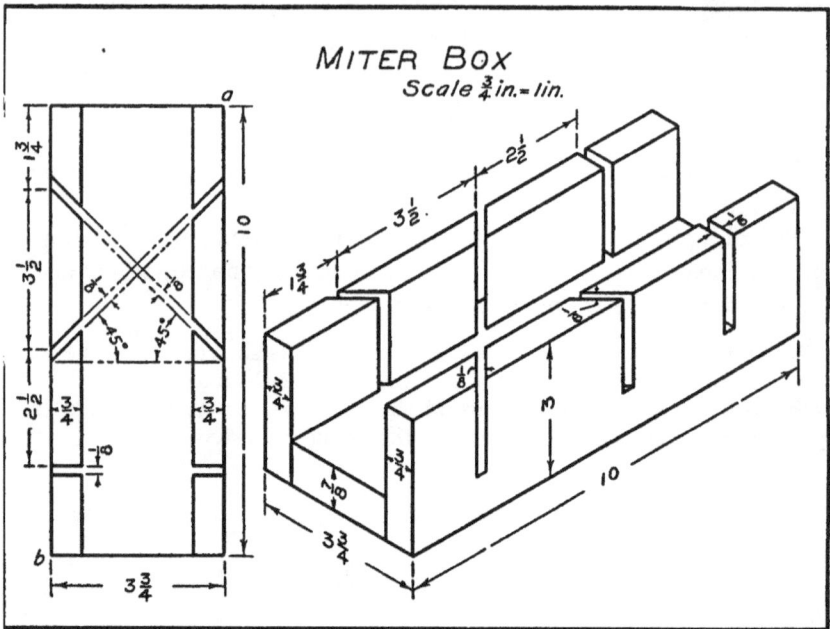

DRAWER AND TABLE JOINTS

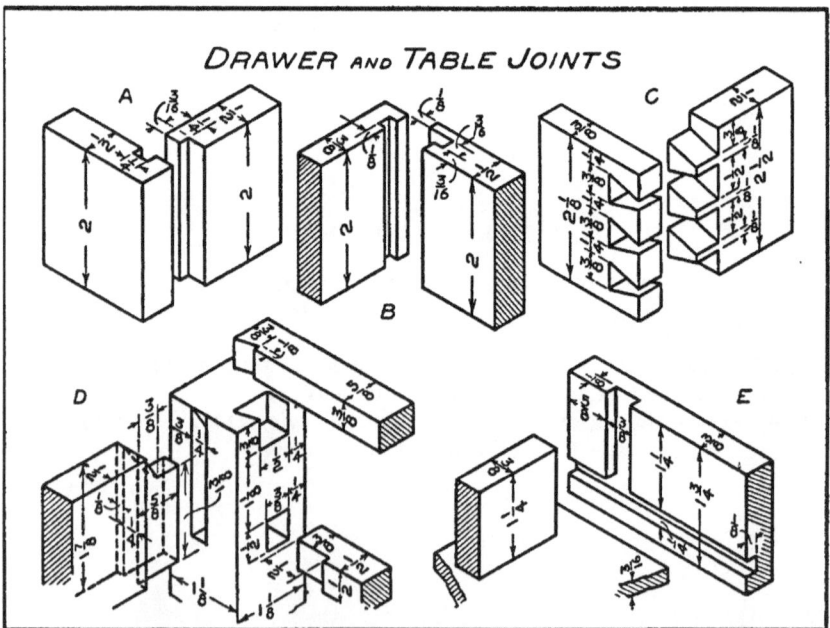

PLATE 72

BOX WITH HINGED LID

This drawing shows the end, and isometric views, of a small box with hinged lid.

Since the lid is partly opened, it will be necessary to draw non-isometric lines. These lines are drawn from points located by *offsets*. For illustration: To locate c', lay off $a'b'$ equal to ab, and on a vertical line drawn from b' lay off bc, giving c'. Draw a line from c' to f'. In a similar manner locate e' and draw $c'e'$. To complete the cover, see Principle 4, page 96.

The end view may be drawn as shown for the following problems.

Problem 1.—Draw the end view as shown, and make an isometric drawing with the corner m at the origin of the axes. Omit hidden lines.

Problem 2.—Draw the end view as shown, and make an isometric drawing with the corner n at the origin of the axes. Omit hidden lines.

SAWHORSE

This drawing shows an isometric construction drawing, and an isometric view of a sawhorse.

The method for finding the offsets required for drawing the legs is shown in the construction drawing and is as follows: Draw the isometric axes and, using dimensions obtained from the isometric drawing, lay off bg and gh. From these points draw lines parallel to the inclined axes. From b lay off distances for c, d, f, and e. Let bf and fe equal 7, and $1\frac{1}{4}$ inches, respectively. The intersections of lines drawn parallel to the inclined axes from these points will give the upper end of one leg, which is completed by drawing lines to the lower end, previously found. A study of the construction drawing should enable the student to complete the isometric drawing.

Problem 1.—Make a construction drawing, as shown, and an isometric drawing, of the sawhorse with the point x at the origin of the axes.

Problem 2.—Make a construction drawing, as shown, and an isometric drawing, of the sawhorse with the point y at the origin of the axes.

PLATE 73

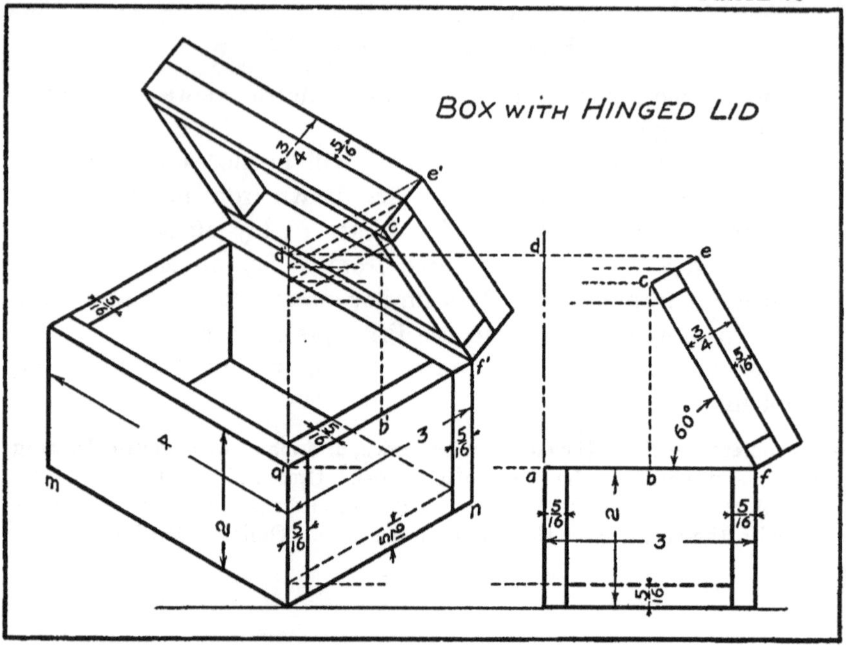

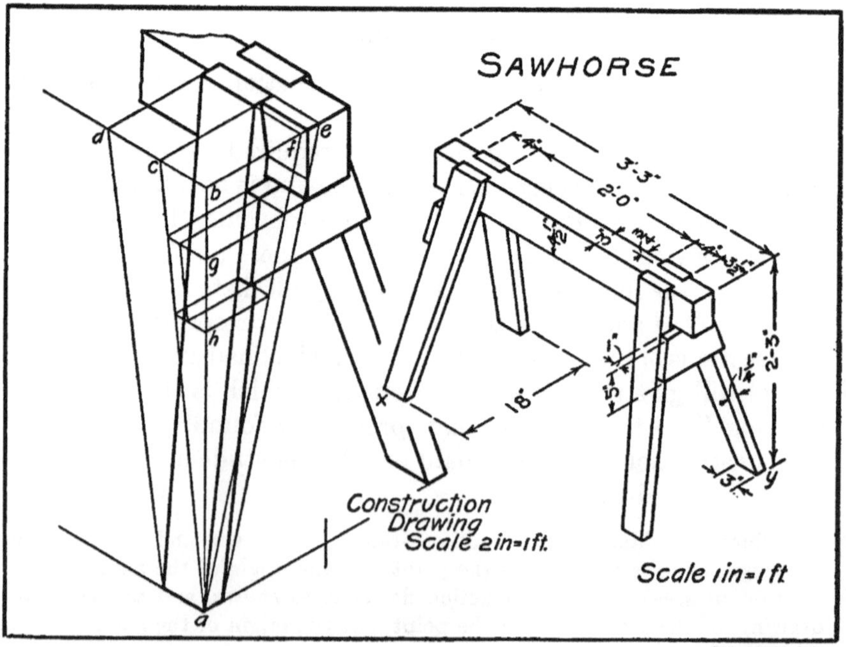

BOX WITH HINGED LID

3/4 5/16

e'

c'

d

e

c

d

f'

60°

3/4 5/16

b

4 3 5/16

a b f

m

2 5/16

5/16

n

5/16 2 5/16

3

5/16

PLATE 74

SAWHORSE

d

c

f e

b

g

h

a

3'-3"

2'-0"

2'-3"

18"

5"

3"

x

y

Construction
Drawing
Scale 2 in = 1 ft.

Scale 1 in = 1 ft

ISOMETRIC PRISMS

This drawing shows the isometric views of four right prisms located from the intersections of their axes as centers.

To make an isometric view of a triangular prism, proceed as follows: Draw the top view of the prism and the axes xx and yy. Then draw the isometric axes xx and yy and lay off distances a, b, and c, as found from the top view, and draw the base of the prism as shown. Draw the vertical edges, measure the height on one edge, and complete the drawing by drawing lines parallel to those of the base.

Problem 1.—Make a drawing of the prisms as shown. Let the bases be regular polygons, the sides of which are $1\frac{1}{2}$, $\frac{7}{8}$, $\frac{3}{4}$, and $\frac{5}{8}$ inches, respectively. Assume a suitable height for the figures.

Problem 2.—Make a drawing showing the prisms when turned in a counterclockwise direction through an angle of 90° about their central vertical axes. Assume suitable dimensions.

PRISMS

This drawing shows the top view and the front view, also the isometric drawing, of a pentagonal prism resting on an edge of an hexagonal prism; the pentagonal prism making any suitable angle with H.

To make an isometric drawing it is necessary to draw the front and the top views and substitute actual dimensions instead of the letters. From a point o draw isometric axes and lay off the offsets, as shown. A study of the drawings should enable the student to execute the isometric view.

Problem 1.—Make a drawing showing a prism 4 inches long, the base being a regular pentagon of 1-inch sides, which rests on a prism $3\frac{1}{4}$ inches long, the base being a regular hexagon of $\frac{7}{8}$-inch sides. Assume a distance for g, and an angle for θ.

Problem 2.—Make a drawing showing a prism 4 inches long, the base being an equilateral triangle of $1\frac{1}{2}$-inch sides, which rests on a prism $3\frac{1}{4}$ inches long, the base being a regular pentagon of 1-inch sides. Assume a distance for g, and an angle for θ.

PLATE 7

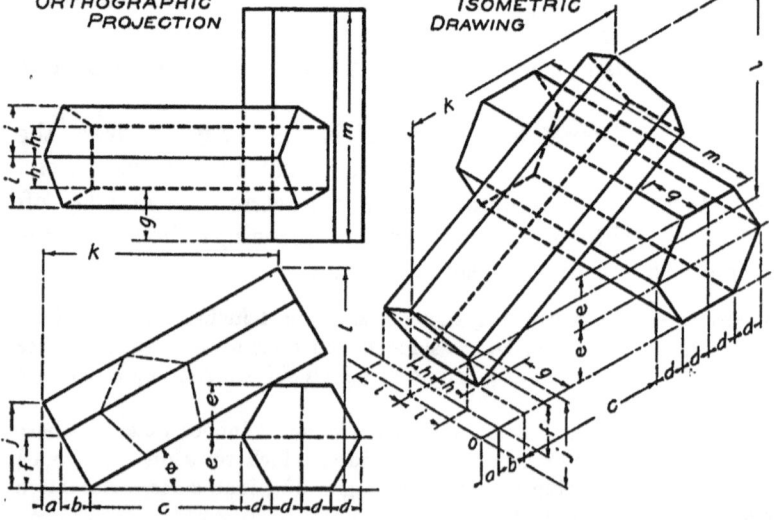

ISOMETRIC PRISMS

TOP VIEWS

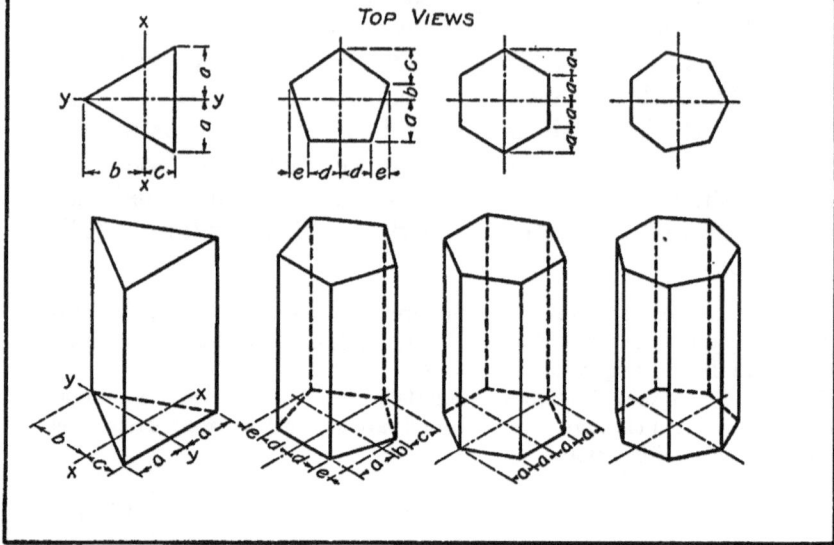

PRISMS

ORTHOGRAPHIC
PROJECTION

ISOMETRIC
DRAWING

PLATE 76

·· ISOMETRIC CIRCLES

To draw a true isometric circle, proceed as follows: Draw a circle showing the required diameter, circumscribe a square, and locate points 1, 2, 3, etc., as shown in *A*. Draw an isometric square, as shown in *B*, and locate points 1, 2, 3, etc., transferred from *A* by measuring the distances parallel to the axes. Through these points draw a smooth curve.

An approximate isometric circle may be drawn from four centers as shown in *D*. Centers 1 and 2 are on the intersections of lines drawn from *a* and *c*, and perpendicular to the opposite sides of the square. This is known as the "four-center" method.

A closer approximation to a true ellipse is shown in *E*. In this drawing eight centers are used. A study of the figure will enable the student to draw the circle. This is known as the "eight-center" method.

Problem 1.—Draw *A*, *B*, *D*, and *E* as shown. Draw *C* similar to *B*, and draw *F* by the four-center method. Assume a suitable diameter.

Problem 2.—Draw *A*, *B*, *D*, and *E* as shown. Draw *C* similar to *B*, and draw *F* by the eight-center method. Assume a suitable diameter.

ISOMETRIC ARCS

This drawing shows approximate isometric arcs drawn by the four-center method for a circle. In drawing a semi-circle, two centers are used, while for a quarter-circle only one center is necessary.

To draw the arcs shown in *A*, proceed as follows: Draw the half-square *fdeag*, and bisect *da*. From *e* and *g*, draw lines perpendicular to *ea* and *ag*. The intersection of these perpendiculars gives the point *c*, the center for arc *cg*. The center for arc *ef*, which lies on *ec*, is found by drawing a line from *f* perpendicular to *df*. Compare the letters in *A*, *B*, and *C* with those in *D* of Isometric Circles.

By dropping perpendicular lines of equal lengths from the centers of arcs *fe*, and *eg*, duplicate arcs may be drawn. These arcs will lie in a plane parallel to the plane in which arcs *fe* and *eg* are drawn. A study of the figures will enable the student to work the following problems.

Problem 1.—Draw the figures as shown. For dimensions, measure the drawing to a scale ⅜ inch equals 1 inch.

Problem 2.—Draw the figures shown, with the corners *x* at the origin of the axes. For dimensions, measure the drawing to a scale ⅜ inch equals 1 inch.

ISOMETRIC CIRCLES

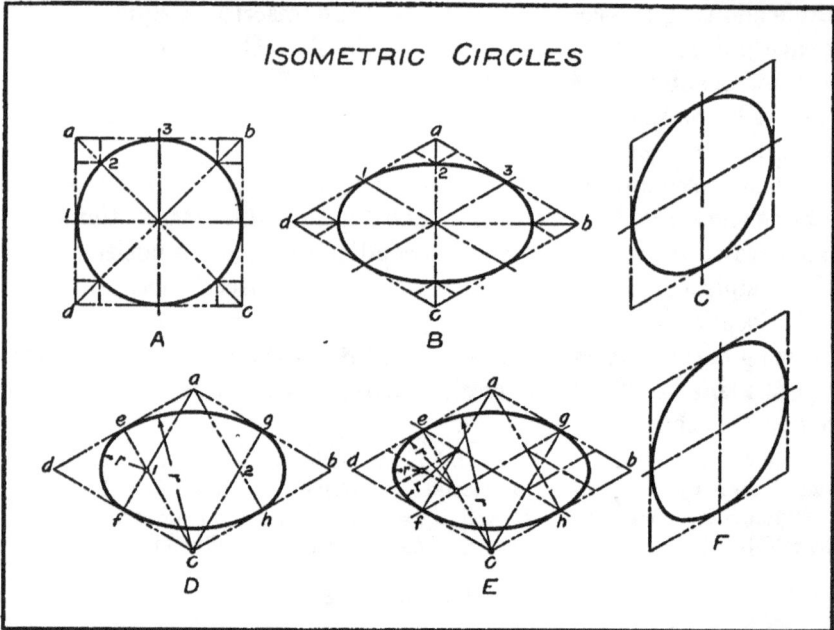

A   B   C

D   E   F

ISOMETRIC ARCS

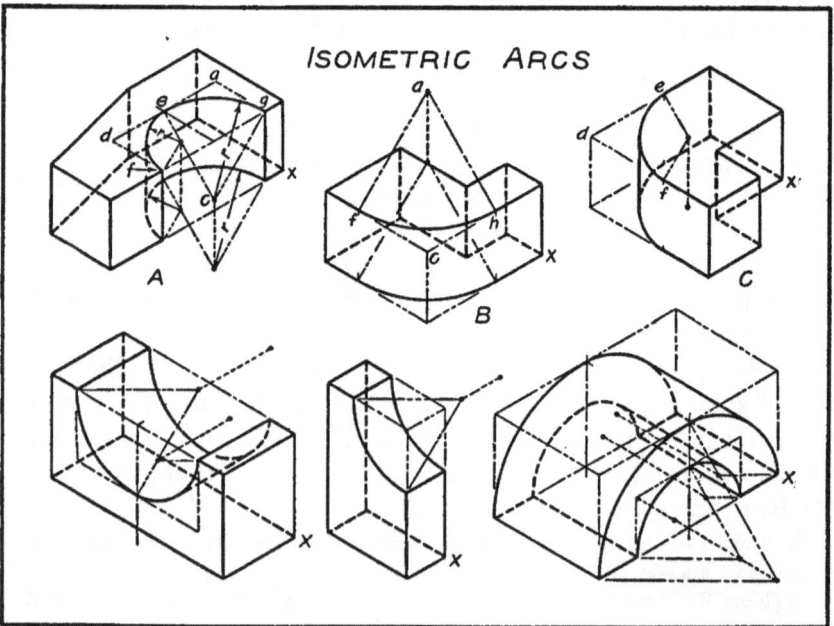

A   B   C

PLATE 78

HOLLOW CYLINDER

This drawing shows two cylinders drawn by the "eight-center" method. The letters in the surface *abcd*, correspond to those shown in *E* of ISOMETRIC CIRCLES. In drawing an object containing concentric circles, as shown in *B*, it is best to erase the construction lines required for one circle before the next is begun.

Problem 1.—Draw the cylinders by the four-center method, of the dimensions shown, with points *c* at the origin of the axes.

Problem 2.—Draw the cylinders by the eight-center method, of the dimensions shown, with points *c* at the origin of the axes.

BEARING CAP

This drawing shows the top and the front views, and the isometric drawing of a bearing cap. It illustrates an object containing concentric arcs; also arcs in parallel planes. The arcs in the vertical planes are quarter-circles. The centers of the arcs can readily be found by referring to the drawings on ISOMETRIC CIRCLES and ISOMETRIC ARCS.

Problem 1.—Draw the front and the top views as shown, and make an isometric drawing with point *x* at the origin of the axes.

Problem 2.—Draw the front and the top views showing the object inverted, and make an isometric drawing with point *y* at the origin of the axes.

PLATE 79

HOLLOW CYLINDER

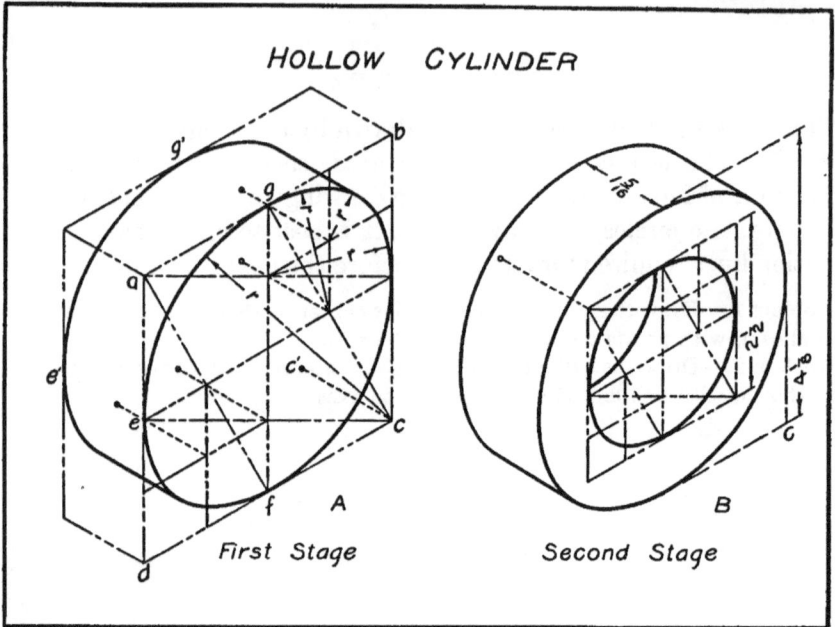

A
First Stage

B
Second Stage

BEARING CAP

ORTHOGRAPHIC PROJECTION

ISOMETRIC DRAWING

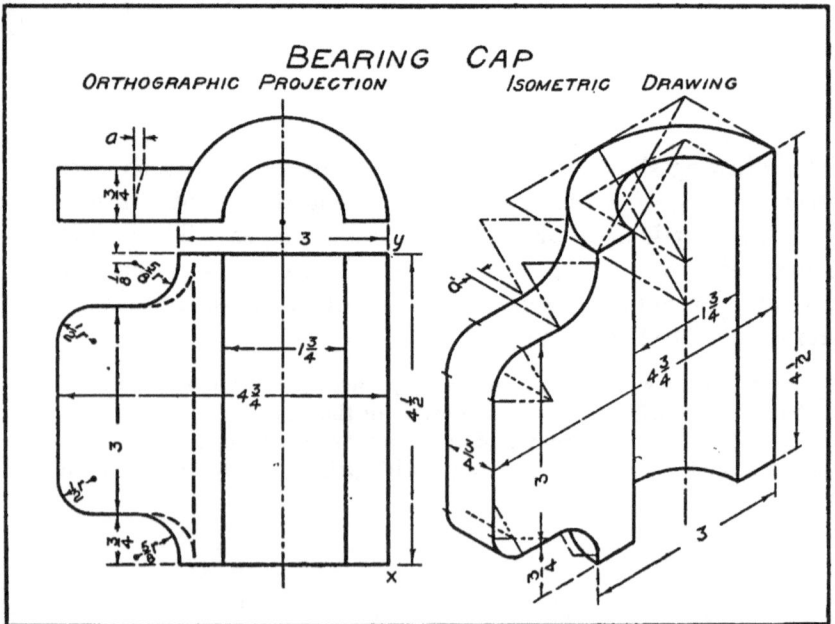

PLATE 80

MAGNET POLE PIECES

This drawing illustrates two objects each containing circular arcs in parallel planes. The object on the left shows arcs in four horizontal planes. The object on the right contains arcs in two parallel vertical planes.

Problem 1.—Make a drawing of the objects with points *a* at the origin of the axes. Show hidden lines for the bore in the figure on the left. Omit construction lines in both figures.

Problem 2.—Make a drawing of the objects with points *b* at the origin of the axes. Show all hidden lines in the figure on the right. Omit construction lines in both figures.

MILLING CUTTER AND FACE PLATE

This drawing shows an isometric view of a milling cutter having sixteen teeth; and a face plate in part section, to show its internal construction.

To divide an isometric circle into any number of equal parts, as in the milling cutter; divide a true half-circle into half the required number of parts and project these divisions to the isometric circle as shown.

To draw the thread in the face plate, draw three-quarter circles showing the beginning and the internal diameter of the thread. On lines drawn from the centers of the arcs found, and parallel to the right axis, space off a series of points at a distance apart equal to the pitch of the thread. (For an explanation of "pitch" refer to drawings of Screw Threads, Bolts, and Nuts.) With these points as centers draw the remaining three-quarter circles.

Problem 1.—Draw a milling cutter with the dimensions shown, and having twenty teeth.

Draw a full face plate with dimensions as shown.

Problem 2.—Draw the milling cutter in a vertical position, showing it revolved about the axis *ab*, and having the dimensions shown.

Draw the face plate showing the face resting on a horizontal plane, and of dimensions shown. Assume a suitable section to show the threads.

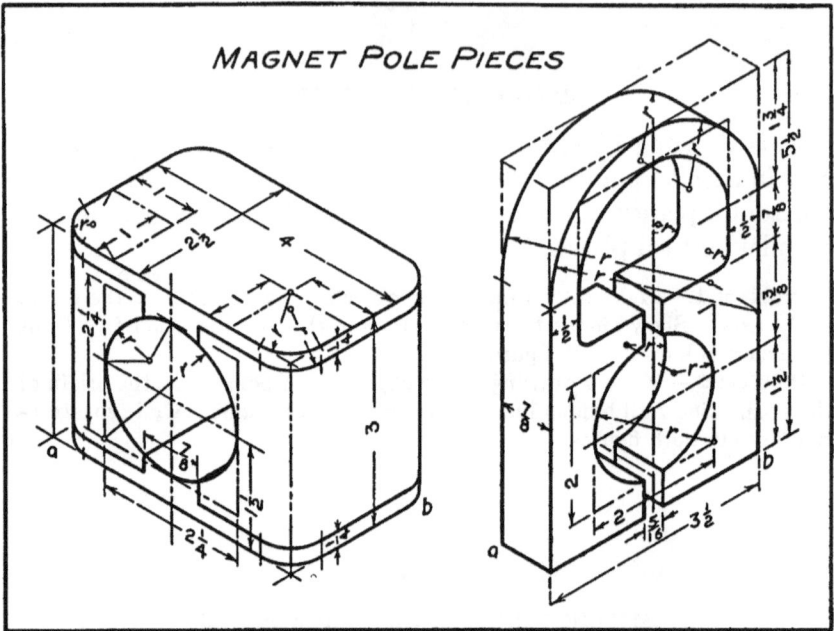

MAGNET POLE PIECES

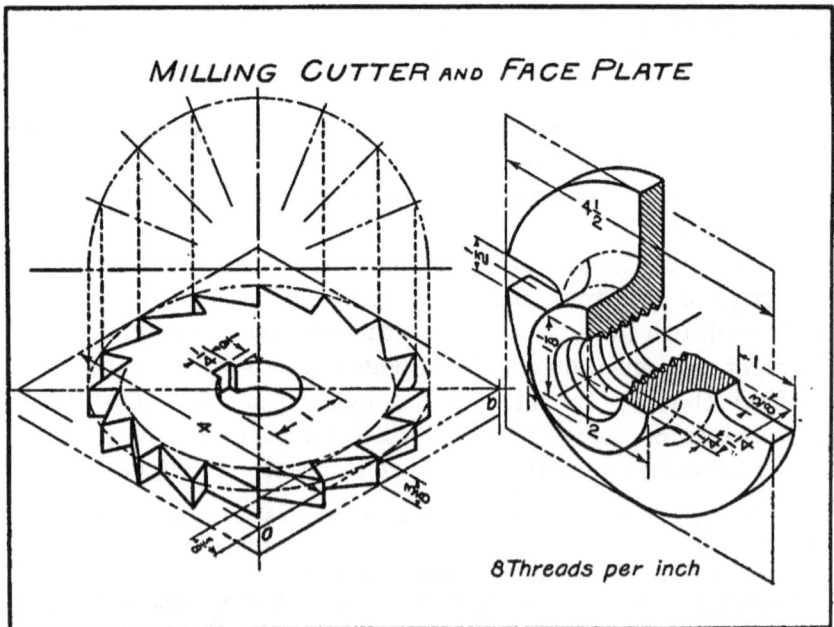

MILLING CUTTER AND FACE PLATE

8 Threads per inch

PLATE 82

KNIFE AND FORK BOX

This drawing shows the method of procedure for finding isometric views of curved lines.

To find curved lines, as shown in *A*, proceed as follows: Draw vertical lines in *A* cutting the curved lines. Draw similar vertical lines in the isometric view, and from line 6–6 lay off the distance 1–1′, 1–1″, 3–3′, 3–3″, etc., equal to the distances found in *A*. Through the points found draw smooth curves. From the intersections of the vertical lines and the curves drawn, draw short lines parallel to the left axis, as shown in 4′–4′, lay off distances equal to the thickness, and draw smooth curves.

Problem 1.—Make a drawing of the box with point *a* at the origin of the axes. Scale 6 in. = 1 ft.

Problem 2.—Make a drawing of the box with point *b* at the origin of the axes. Scale 6 in. = 1 ft.

·

UNIFORM MOTION CAM

This drawing shows the top and the front views, also the isometric drawing of an object containing curved lines, and circles, which lie in parallel vertical planes.

To make an isometric drawing it is necessary to first construct a drawing showing the front and the top views of the object.

The curve *egf*, in the front view, is found by dividing the half-circle, also the line *de*, each into six equal parts and drawing circular arcs as shown. Through the points of intersection of the circular arcs and radial lines draw a smooth curve.

A study of the drawing should enable the student to work the following problems.

Problem 1.—Make a drawing showing the top, front, and isometric views of the object with point *b* at the origin of the axes.

Problem 2.—Make a drawing showing the top, front, and isometric views of the object with point *c* at the origin of the axes.

PLATE 83

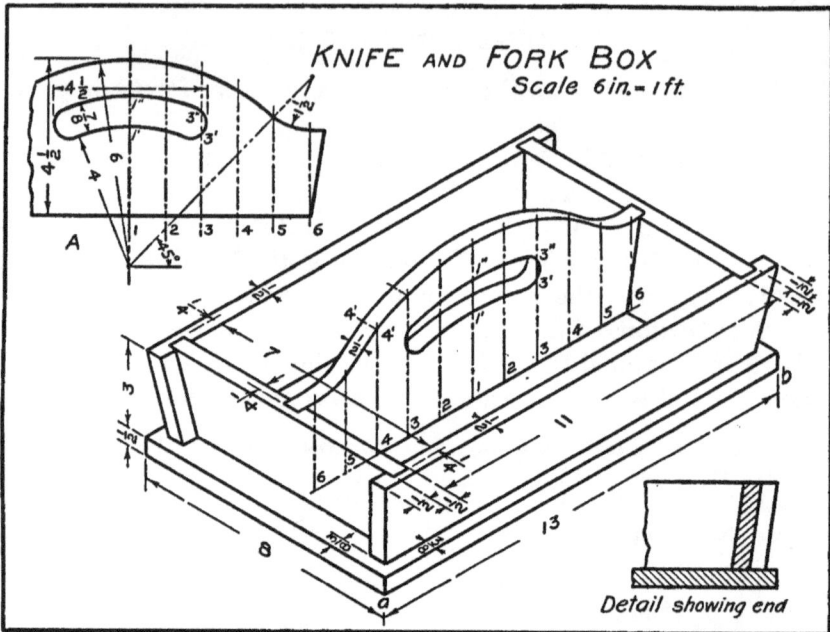

KNIFE AND FORK BOX
Scale 6 in. = 1 ft.

A

Detail showing end

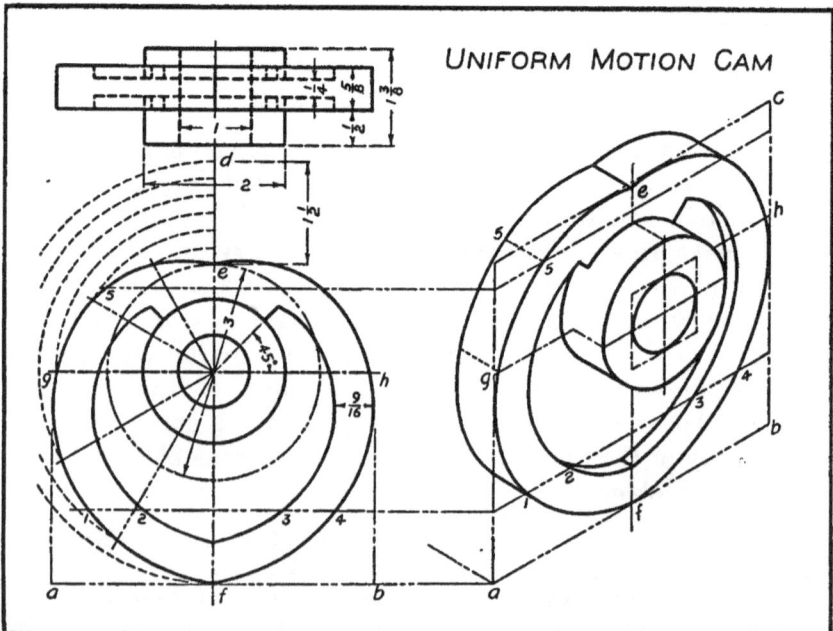

UNIFORM MOTION CAM

PLATE 84

CAVALIER PROJECTION

Cavalier projection, also called oblique projection, is somewhat similar to isometric drawing since it has three axes on which actual lengths are measured.   One axis is horizontal, one vertical, and one oblique.   The oblique axis may make any angle with a horizontal; 30° or 45° are, however, most generally used.

BRACKET SHELF

The drawing shows the object with its oblique axis making an angle of 30° below a horizontal, thus showing a lower, or underneath view.

Since one axis is horizontal and one vertical, the surface of the back is drawn in its true shape, being parallel to the vertical plane of projection.

The circular arcs in the bracket are drawn by the four-center method, and the front edge of the shelf is found by offsets as shown in the detail.

Problem 1.—Make a cavalier projection of the bracket as shown.

Problem 2.—Make a cavalier projection of the bracket showing the front and left faces, instead of the front and right as shown.

CABINET DRAWING

Cabinet drawing, also called oblique drawing, has three axes on which measurements are made.   One axis is horizontal, one vertical, and one oblique.   Actual lengths are measured on the horizontal and the vertical axes, and one-half actual lengths are measured on the oblique axis.

Drawings made by this method show less distortion than do Isometric Drawings or Cavalier Projections.

KITCHEN TABLE

The drawing shows a portion of the table top removed to show the construction more clearly.   It also shows the drawer partly opened.

Problem 1.—Make a cabinet drawing of the table showing the drawer closed.   Scale 2 in. = 1 ft.

Problem 2.—Make a cabinet drawing of the table on axes as shown in *B*, thereby giving a view from underneath.   Show the drawer pulled out 6 inches.   Scale 2 in. = 1 ft.

PLATE 85

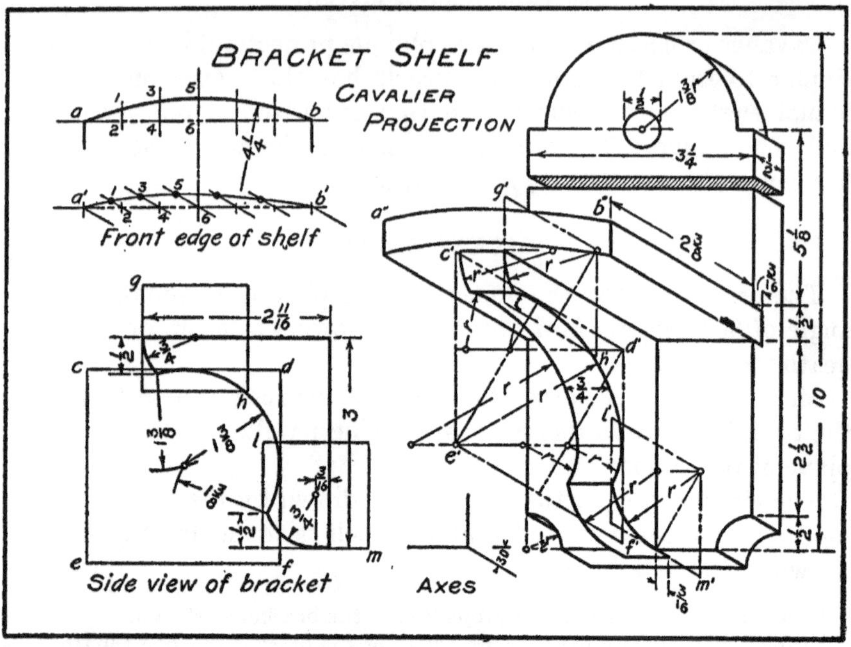
BRACKET SHELF
CAVALIER PROJECTION
Front edge of shelf
Side view of bracket
Axes
30°

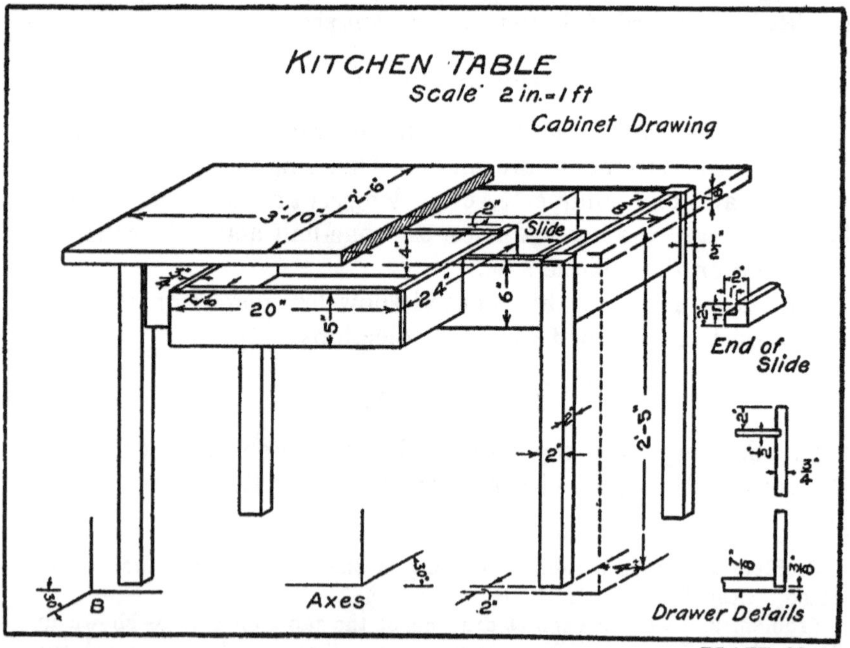
KITCHEN TABLE
Scale 2 in.=1 ft
Cabinet Drawing
3'-10"
2'-6"
2"
Slide
4"
6"
24"
20"
5"
2'-5"
End of Slide
Drawer Details
Axes
30°
B

PLATE 86

KNUCKLE JOINT

This drawing shows an object containing a number of circles and circular arcs in parallel vertical planes; also a number of circular arcs in horizontal planes.

The circles and arcs in the vertical planes are drawn with the compass. The arcs in the horizontal planes must be drawn through points plotted from the top view shown below.

Problem 1.—Draw a top view and make a cabinet drawing of the object as shown.

Problem 2.—Draw a top view and make an assembly cabinet drawing of the object showing the bolt in place.

SMALL BENCH

An object may be revolved into any number of positions and still show three faces.

The object shown in the drawing is revolved to a position so that the axes make angles as shown in the axes diagram.

The arches shown in the legs are half-circles and may be drawn as shown by the two figures on the left; while the quarter-circles in the top are isometric quarter-circles.

Problem 1.—Make a drawing showing the object with *a* at the origin of the axes. Let the left axis make an angle of 45°, and the right axis an angle of 15°, with a horizontal.

Problem 2.—Make a drawing showing the object with *b* at the origin of the axes. Let the left axis make an angle of 7°, and the right axis an angle of 41°, with a horizontal. Measure actual lengths on the left and the vertical axes, and one-half actual lengths on the right axis.

Note.—Since the width of the object is reduced to half the actual dimensions the arcs in legs and top cannot be drawn with the compass but must be plotted by offsets, as shown in the drawing of the KNIFE AND FORK Box.

PLATE 87

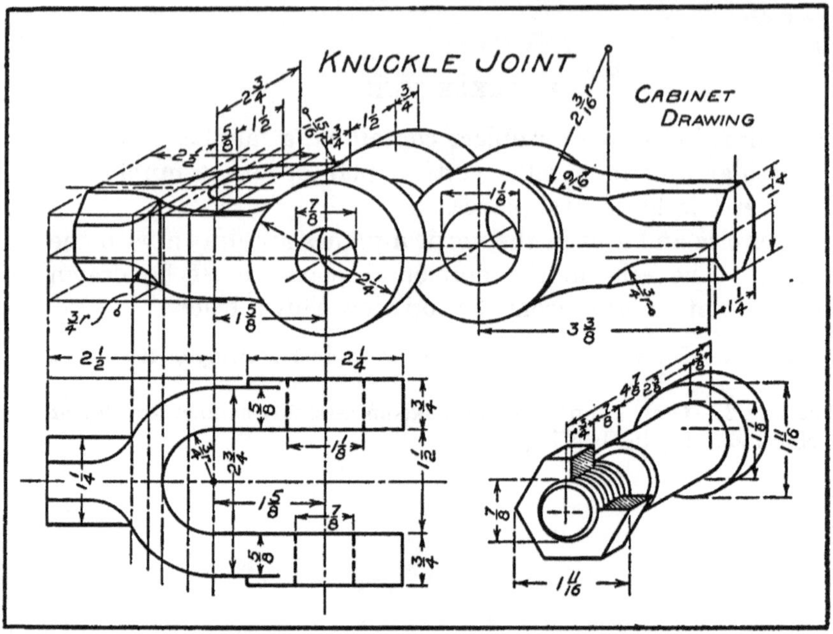

KNUCKLE JOINT

CABINET DRAWING

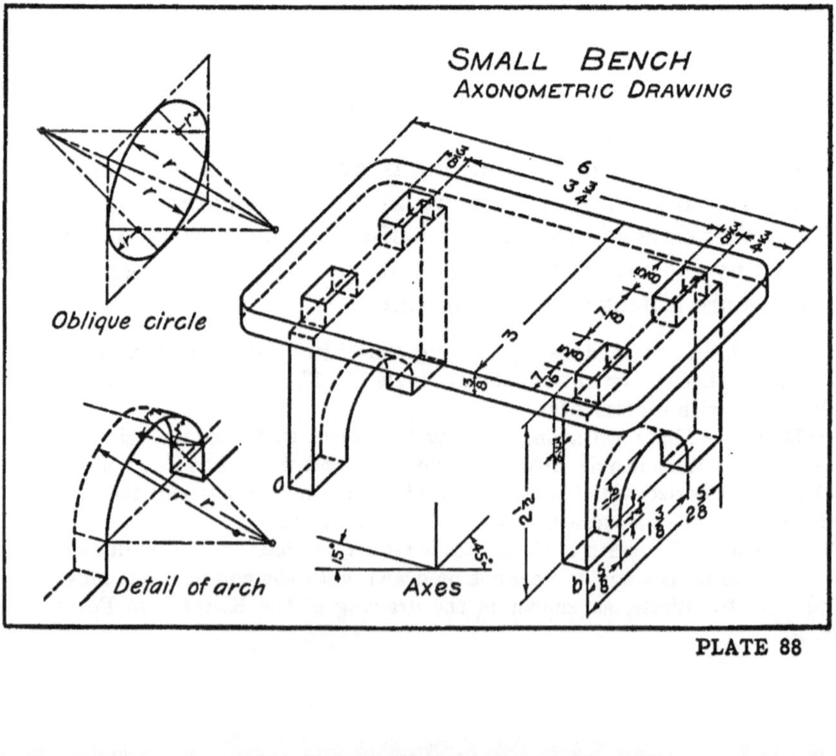

SMALL BENCH
AXONOMETRIC DRAWING

Oblique circle

Detail of arch

Axes

PLATE 88

SECTION IV

MACHINE DETAILS

Cast Iron
ECCENTRIC SHEAVE

The drawing shows a front view and an incomplete side view. The latter is to be completed when working the problems. The distance a is the eccentricity, or throw, and is equal to one-half the valve travel.

The following proportions are for eccentrics to 5 inches valve travel:

D = diameter of shaft	$f = D + \frac{1}{16}$	$l = e - \frac{1}{2}$
a = throw	$g = 2\frac{3}{4}D - \frac{1}{16}$	$m = 2e - \frac{1}{16}$
$b = 1\frac{1}{2}D$	$h = \frac{1}{2}e + \frac{1}{16}$	$n = \frac{1}{2}e - \frac{1}{4}$
$c = 2\frac{1}{2}D$	$j = \frac{1}{2}e - \frac{3}{8}$	$o = \frac{1}{2}e + \frac{7}{64}$
$e = \frac{1}{2}D + \frac{3}{8}$	$k = e - \frac{5}{16}$	$p = \frac{1}{4}e - \frac{5}{32}$

Problem 1.—Make a drawing of an eccentric sheave for a 2-inch shaft and $\frac{3}{4}$-inch throw. Draw full size.

Problem 2.—Make a drawing of an eccentric sheave for a 4½-inch shaft and 2½-inch throw. Scale 6 in. = 1 ft.

Problem 3.—Assume a value for shaft diameter and a throw for an eccentric sheave, and draw two views.

Machine
HAND-WHEEL

The drawing shows a full cross-section and an incomplete front view. The latter is to be completed when working the problems.

To construct an arm, draw straight line 5–6 giving point 7. Through point 7 draw a line making angle 6–7–8 equal to angle 7–6–8, giving centers 8 and 9. From these centers the center-line of the arm may be drawn. The centers for the arcs, giving the thickness of the arm, must lie on line 9–8, and are to be found by trial. Omit all construction lines in the drawing when inking.

The following proportions are for wheels up to 12 inches diameter:

A = diameter of wheel	$e = \frac{1}{8}A - \frac{1}{16}$	$i = \frac{3}{32}A + \frac{9}{32}$
$b = \frac{3}{32}A + \frac{13}{32}$	$f = \frac{3}{32}A + \frac{13}{32}$	$j = \frac{1}{16}A + \frac{1}{16}$
$c = \frac{1}{16}A + \frac{1}{16}$	$g = \frac{1}{2}b - \frac{1}{16}$	$k = \frac{1}{2}h$
$d = \frac{1}{16}A + \frac{1}{4}$	$h = 2d - \frac{5}{8}$	$l = \frac{1}{2}j$

Problem 1.—Make a drawing for a 5-inch hand-wheel with four arms. Draw full size.

Problem 2.—Make a drawing for a 10-inch hand-wheel with six arms. Scale 6 in. = 1 ft.

Problem 3.—Assume a diameter for a hand-wheel and draw two views.

PLATE 89

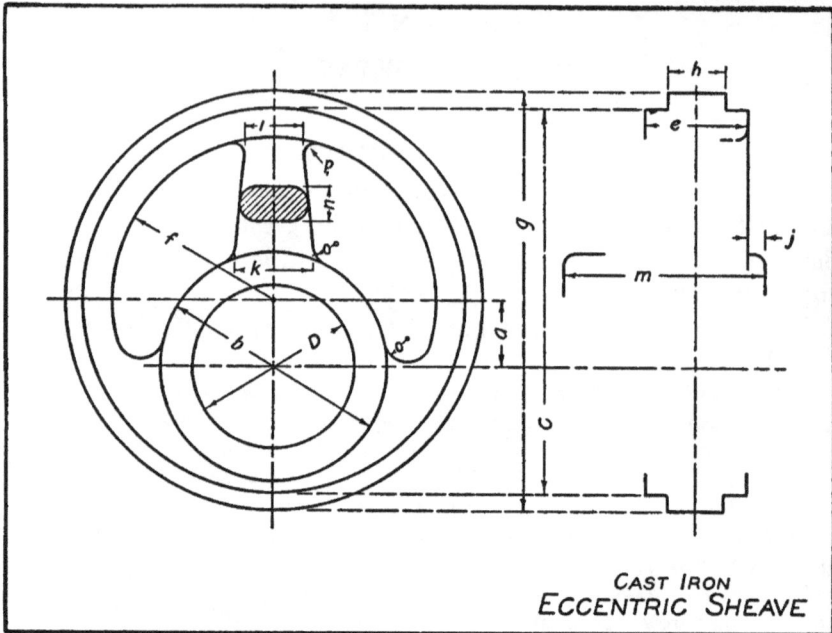

CAST IRON
ECCENTRIC SHEAVE

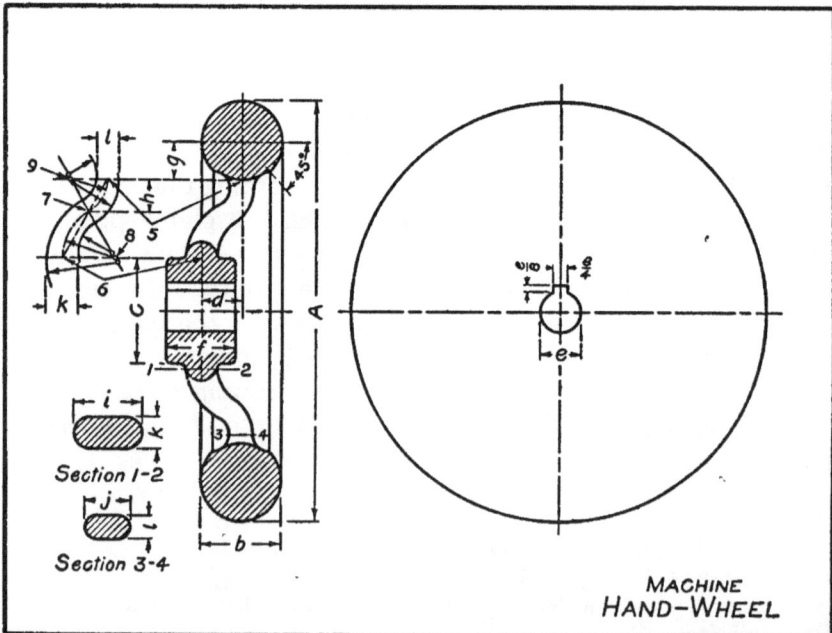

Section 1-2

Section 3-4

MACHINE
HAND-WHEEL

PLATE 90

Design of
ENGINE CRANK

Cranks, like eccentrics, are devices used for transforming rotary motions into reciprocating motions. They are sometimes made of cast iron, but usually of cast steel or machine steel.

The lines of intersection in the side view are found by assuming points on the fillets, projecting to the front view and back to the side view, as indicated by the direction of the arrows in the lower curve.

A = throw of crank $f = \frac{1}{2}D + \frac{1}{16}$ $l = \frac{1}{2}D + \frac{1}{2}$

D = diameter of shaft $g = D$ $m = D - \frac{1}{4}$

$b = 1\frac{3}{4}D$ $h = \frac{9}{16}D + \frac{1}{8}$ $n = \frac{1}{4}D$

$c = 1\frac{1}{8}D + \frac{1}{8}$ $j = \frac{11}{16}D + \frac{1}{4}$ $p = \frac{1}{4}D + \frac{1}{16}$

$e = \frac{1}{2}D + \frac{1}{4}$ $k = \frac{7}{16}D + \frac{3}{32}$

Problem 1.—Make a drawing of a crank for $4\frac{1}{2}$-inch throw and $1\frac{1}{2}$-inch shaft. Draw full size.

Problem 2.—Make a drawing of a crank for 7-inch throw and $2\frac{1}{2}$-inch shaft. Scale 6 in. = 1 ft.

Problem 3.—Assume values for A and D and draw two views of a crank.

Machine Steel
CONNECTING-ROD END

Connecting-rods are used in steam engines to join the crank with the cross-head; also in gasoline engines to connect the piston with the crank shaft. They convert the reciprocating motion of the piston to a rotary motion of the shaft. Connecting-rods are made of machine steel, although for some types of small gasoline engines they may be made of bronze.

The drawing shows the cross-head end, or piston end, of a solid rod. The hole, when the rod is made of steel, is lined with white metal or bronze. This lining is called a "bushing," and is frequently made about $\frac{1}{8}$ inch thick.

D = outside diameter of bushing $c = 1\frac{3}{4}D + \frac{7}{32}$

$a = 1\frac{1}{4}D + \frac{1}{32}$ $d = 2D + \frac{1}{4}$

$b = D + \frac{1}{8}$ $e = D - \frac{3}{8}$

Problem 1.—Make the drawing for a connecting-rod end when the outside diameter of the bushing is $1\frac{3}{4}$ inches. Draw full size.

Problem 2.—Make the drawing for a connecting-rod end when the outside diameter of the bushing is 3 inches. Scale 6 in. = 1 ft.

Problem 3.—Assume a value for D and draw three views of a connecting-rod end.

PLATE 91

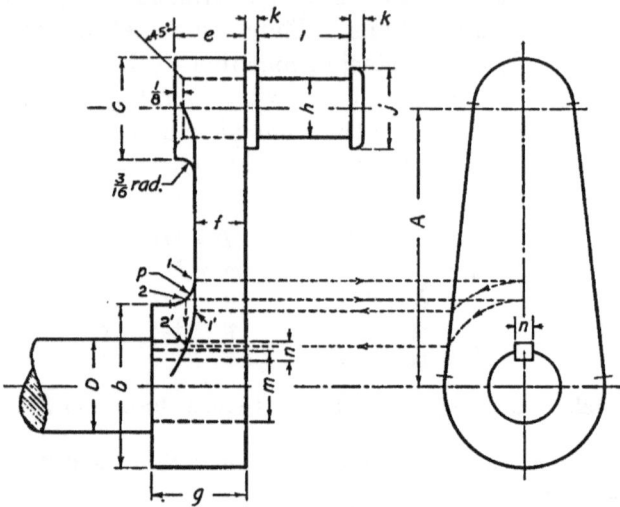

DESIGN OF
ENGINE CRANK

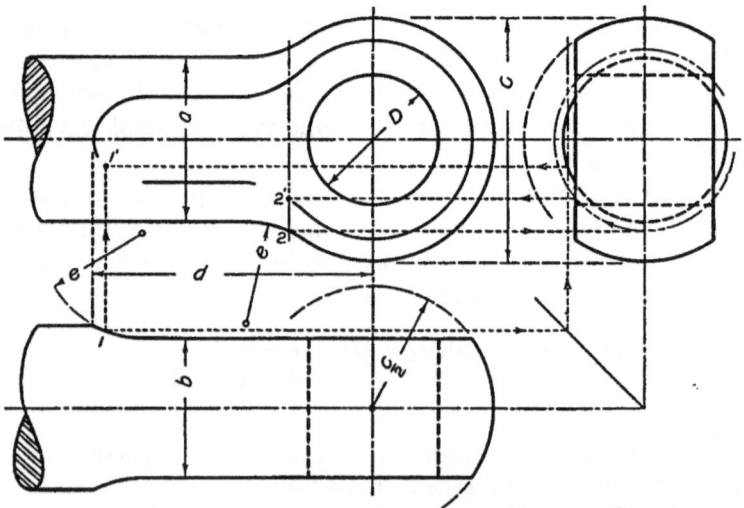

MACHINE STEEL
CONNECTING-ROD END

PLATE 92

Soft Steel
CRANE HOOK

Cranes are machines used for hoisting or lowering weights, and crane hooks are attached to these machines by means of a steel cable or rope for holding or sustaining the weights. The hooks are generally made of soft steel.

The drawing shows the front view, the side view, and three cross-sections of a hook.

a = throat opening	f = .7 a	k = .85 a
b = 5.4 a	g = 1.2 a	l = .9 a
c = 1.5 a	h = .85 a	m = .2 a
d = 1.5 a	i = .1 a	n = .6 a
e = .2 a	j = .8 a	o = .8 a

Problem 1.—Make a drawing showing two views and sections for a crane hook having a $1\frac{1}{4}$-inch throat opening. Draw full size.

Problem 2.—Make a drawing showing two views and sections for a crane hook having a $2\frac{1}{4}$-inch throat opening. Scale 6 in. = 1 ft.

Problem 3.—Assume the throat opening for a crane hook and draw two views and sections, as shown.

Cast Iron
CLUTCH COUPLINGS

Clutch couplings are used for connecting, or for disconnecting, two shafts. They are generally made of cast iron.

The drawing shows a square jaw coupling, also a spiral jaw coupling, each having four jaws. The left hand portions are keyed to the shafts; the right hand portions are movable and are prevented from turning on the shafts by feather keys. The end views are shown incomplete in the drawing.

D = diameter of shafts	e = $1\frac{1}{2}D + \frac{1}{2}$	j = $\frac{3}{16}D + \frac{3}{16}$
a = $2D + 1$	f = $\frac{1}{2}D$	k = $\frac{5}{16}D + \frac{7}{16}$
b = $1\frac{3}{4}D + 1\frac{3}{8}$	g = $\frac{3}{8}D + \frac{3}{8}$	l = $g + \frac{1}{4}$
c = $1\frac{1}{2}D + \frac{3}{4}$	h = $1\frac{3}{4}D + \frac{7}{8}$	m = $1\frac{1}{4}D + \frac{1}{4}$

Problem 1.—Make a drawing showing front and end views of a square jaw coupling, also of a spiral jaw coupling, each having four jaws. Let D = 1 inch. Draw full size.

Problem 2.—Make a drawing showing front and end views of a square jaw coupling, also a spiral jaw coupling, each having three jaws. Let D = $1\frac{3}{4}$ inches. Show each in half section. Scale 6 in. = 1 ft.

Problem 3.—Assume a diameter for D and draw two views of a square jaw coupling, also a spiral jaw coupling, each having four jaws.

PLATE 93

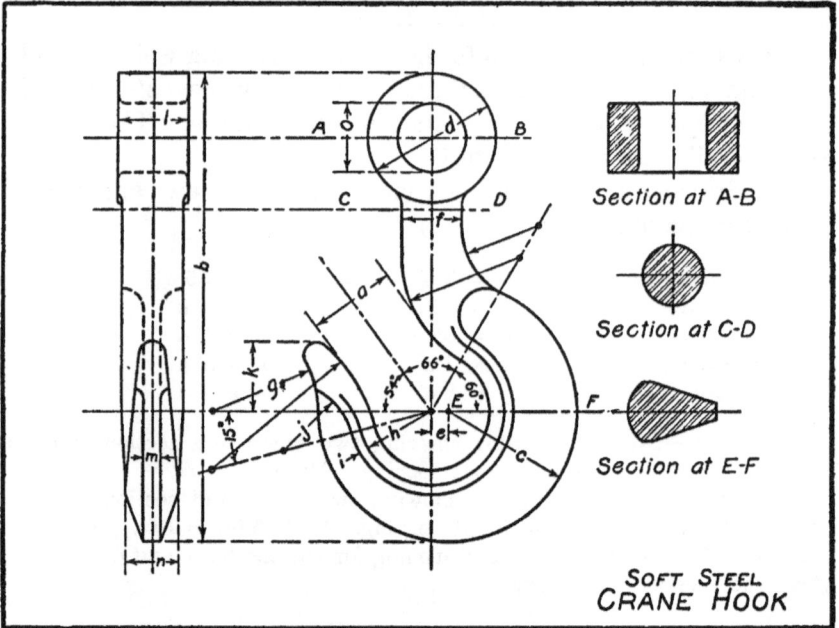

Section at A-B

Section at C-D

Section at E-F

SOFT STEEL
CRANE HOOK

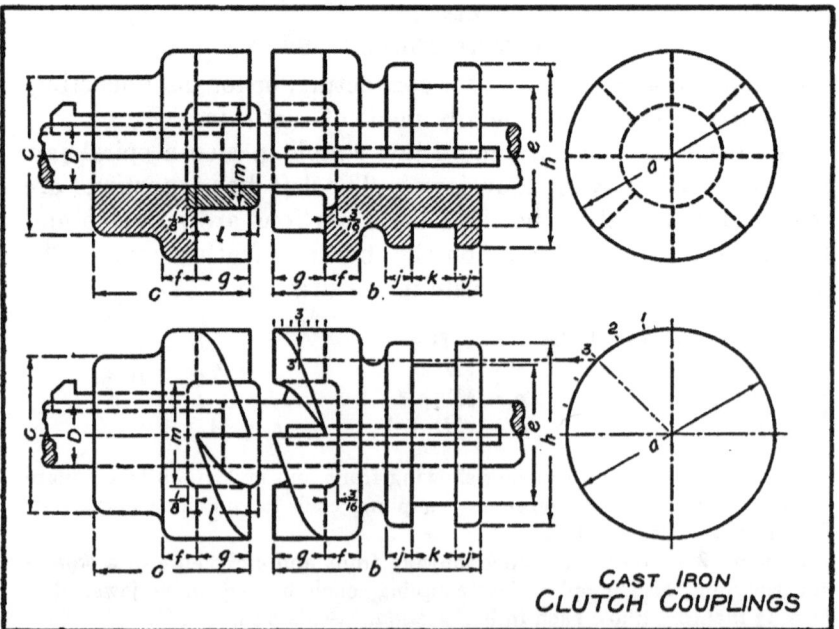

CAST IRON
CLUTCH COUPLINGS

PLATE 94

Screw Threads, Bolts and Nuts

Screws, bolts, studs, and nuts are used in all machines to a greater or less extent. A knowledge of their proportions and their conventional representation is a prime requisite of draftsmen engaged in machine drawing.

The exact representation of a screw thread is a laborious and time-consuming operation, since the curve of a thread is a helix which is plotted by projecting points from an end view. In practice it is not often necessary to draw exact threads; therefore, to economize in time the threads are conventionalized and shown by straight lines.

The distance between two successive threads on a screw is called the "pitch;" that is, the pitch is the distance a screw will advance in the direction of its axis in one revolution.

The drawings show accepted conventions for representing screw threads. The conventions in the upper drawing are suitable for screws of three-quarter inch and less in diameter. The lower drawing shows conventions for screws of larger diameters.

In drawing a conventional screw it is not necessary to show the exact number of threads per inch. The diameter fixes the number of threads, since screws are made "standard," and definite diameters have a definite number of threads per inch. If a non-standard screw is necessary in a machine, the number of threads per inch should be stated in a note on the drawing.

Consult the tables on screws, bolts, and nuts for proportions, and see Figs. 33 and 34 for drawing threads.

Conventions for
SCREW THREADS

Problem 1.—Make a drawing as shown. Assume suitable diameters for d.
Problem 2.—Make a drawing as shown. Let the diameters equal $\frac{1}{8}$, $\frac{3}{4}$, $\frac{5}{8}$, $\frac{1}{2}$, $\frac{7}{16}$, and $\frac{3}{8}$ inches, respectively.

Conventions for
BOLTS AND NUTS

Problem 1.—Make a drawing as shown. Assume suitable diameters for d.
Problem 2.—Make a drawing as shown. Let the diameters equal 1, $\frac{7}{8}$, $\frac{3}{4}$, $\frac{5}{8}$, and $\frac{1}{2}$ inches, respectively.

PLATE 95

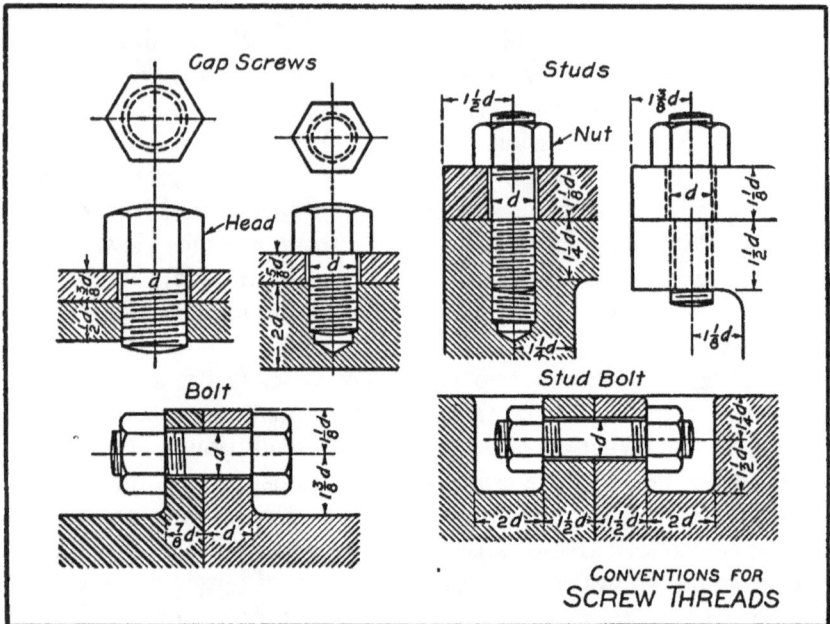

Cap Screws

Studs

Head

Nut

Bolt

Stud Bolt

CONVENTIONS FOR
SCREW THREADS

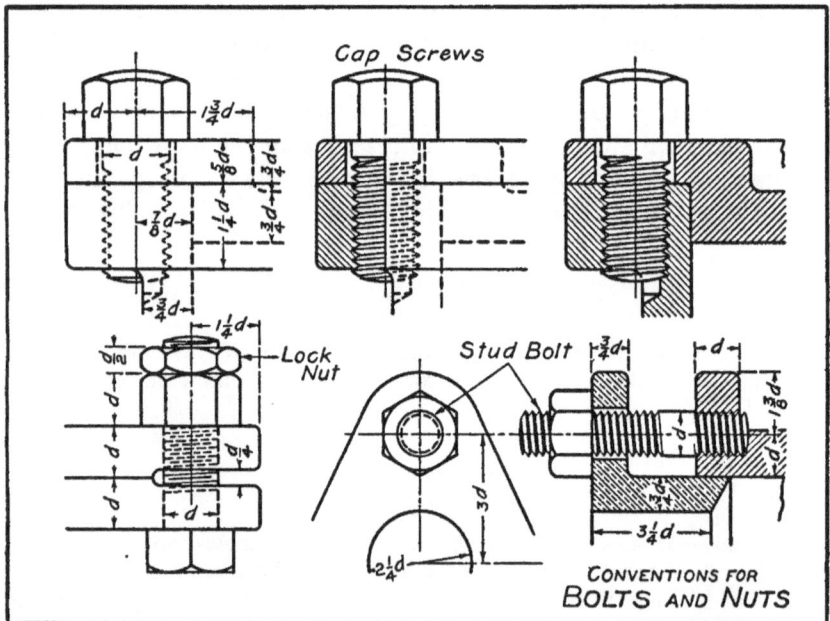

Cap Screws

Lock
Nut

Stud Bolt

CONVENTIONS FOR
BOLTS AND NUTS

PLATE 96

Drop Forged
LATHE CARRIER

Lathe carriers, frequently called lathe dogs, are tools used for driving mandrels in engine lathes. The mandrel supported by the lathe centers is driven by the carrier, which is fixed to one end by a set screw. The carrier is driven by the face plate of the lathe. Lathe carriers may be made of cast iron, cast steel, or drop forgings.

The drawing shows the top view, the end view, and the center line for the front view. The shape of the curve in the top view, on which two points are shown by 4–4, may be found as in previous problems.

A = diameter of opening $e = \frac{1}{16}A$ $j = \frac{1}{2}A - \frac{3}{16}$

$b = 4A - \frac{1}{2}$ $f = \frac{1}{4}A - \frac{5}{32}$ $k = \frac{1}{4}A + \frac{1}{32}$

$c = 1\frac{1}{4}A - \frac{3}{32}$ $g = A - \frac{1}{4}$ $l = \frac{3}{8}A - \frac{5}{64}$

$d = 1\frac{1}{2}A - \frac{5}{16}$ $h = \frac{1}{2}A + \frac{1}{16}$ $m = \frac{3}{8}A$

Problem 1.—Make a drawing showing three views of a carrier when A equals $1\frac{3}{4}$ inches. Draw full size.

Problem 2.—Make a drawing showing three views of a carrier when $A = 3$ inches. Scale 6 in. = 1 ft.

Problem 3.—Assume a diameter for A and draw three views of a carrier.

Cast Iron
CLAMP COUPLING

Clamp couplings, also called split couplings, are used for connecting the ends of shafts. There are many forms of couplings; the one shown is generally used in cramped places. They are almost always made of cast iron.

The coupling in the drawing is provided with a keyway. The insertion of keys prevents the shafts from turning in the coupling. The views are shown incomplete.

D = diameter of shaft $e = \frac{1}{8}b + \frac{5}{32}$ $l = \frac{1}{4}D + \frac{3}{16}$

$a = 2D + 3\frac{1}{2}$ $f = \frac{1}{4}e - \frac{5}{64}$ $m = 3\frac{1}{2}l + \frac{1}{32}$

$b = 2D + \frac{3}{4}$ $g = \frac{1}{8}b + \frac{5}{16}$ $n = \frac{1}{2}D$

$c = \frac{1}{4}a$ $h = 1\frac{1}{2}g - \frac{11}{64}$ $o = \frac{1}{4}n + \frac{1}{32}$

$d = a - 2c$ $k = l + \frac{1}{16}$ $p = \frac{3}{32}a + \frac{3}{64}$

Problem 1.—Make a drawing showing three complete views for a $1\frac{1}{4}$-inch coupling containing four bolts. Draw full size.

Problem 2.—Make a drawing showing three complete views for a $2\frac{3}{8}$-inch coupling containing four bolts. Scale 6 in. = 1 ft.

Problem 3.—Assume a shaft diameter for a clamp coupling and draw three views.

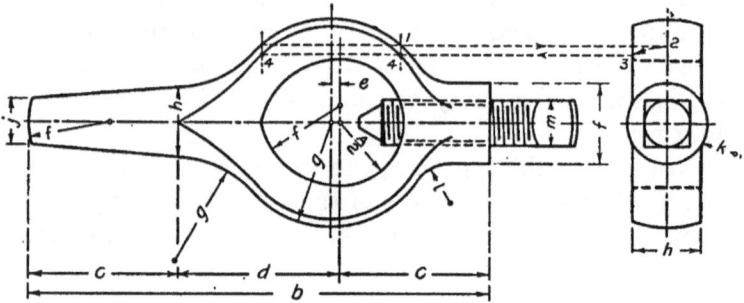

DROP FORGED
LATHE CARRIER

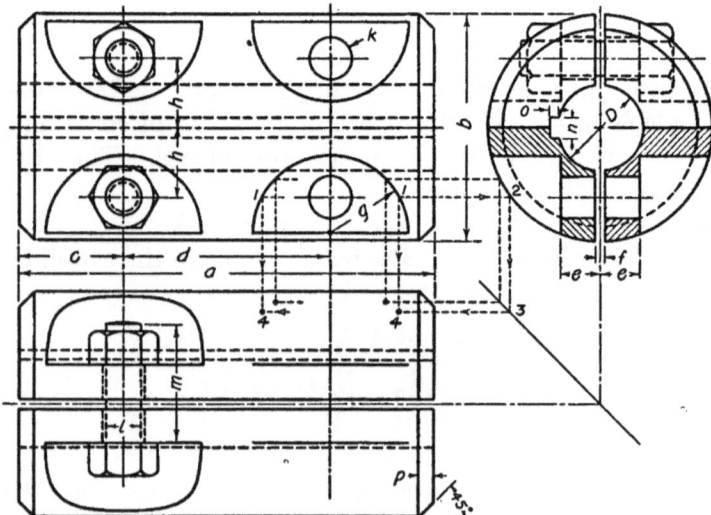

CAST IRON
CLAMP COUPLING

PLATE 98

Piston Rod
STUFFING BOX

Stuffing boxes are used on engines, pumps, and many other machines, to prevent leakage of steam, water, etc. They are usually made of cast iron with two or more parts of bronze encircling the movable rod.

The drawing shows a stuffing box suitable for steam engines. The outer part, or box, is cast iron, while the two parts, called glands, encircling the rod, are bronze. Leakage is prevented by filling the cavity a with an elastic substance which is pressed against the rod by the outer gland. The outer gland is forced into the box by screwing down the nuts.

D = diameter of rod	$g = \frac{11}{16}D + 1\frac{5}{8}$	$n = \frac{11}{16}D + \frac{11}{16}$
$a = D + 1$	$h = \frac{7}{16}D + \frac{5}{16}$	$o = 1\frac{5}{8}D + 1\frac{3}{4}$
$b = D + 1\frac{13}{16}$	$j = \frac{1}{8}D + 1\frac{11}{16}$	$p = \frac{1}{4}D + \frac{3}{16}$
$c = D + \frac{9}{16}$	$k = D + \frac{7}{8}$	$q = 1\frac{1}{2}D + 1\frac{1}{2}$
$e = 1\frac{1}{2}D + 1\frac{11}{16}$	$l = \frac{1}{4}D + \frac{1}{8}$	$r = \frac{3}{16}D + \frac{7}{16}$
$f = \frac{7}{8}D + 2\frac{1}{8}$	$m = \frac{1}{4}D + \frac{3}{16}$	

Problem 1.—Draw front and end views of a stuffing box for a $\frac{7}{8}$-inch rod. Show the front view in half-section. Draw full size.

Problem 2.—Draw two views of a stuffing box for a 2-inch rod. Show the front view in full-section. Scale 6 in. = 1 ft.

Problem 3.—Assume a diameter for D and draw two views of a stuffing box.

Flanged
SAFETY COUPLING

Couplings are frequently provided with a flange or rim at their outer edge to prevent accidents which might occur by a belt or a person's clothing becoming entangled with the bolts. Couplings of this type are made of cast iron.

D = diameter of shaft	$f = \frac{3}{4}a$	$o = 3n + \frac{3}{16}$
$a = 4D$	$g = \frac{1}{64}a + \frac{7}{32}$	$p = \frac{1}{4}D$
$b = 2\frac{1}{2}D - \frac{1}{2}$	$h = \frac{1}{64}a + \frac{7}{32}$	$q = \frac{1}{2}p + \frac{1}{8}$
$c = 1\frac{1}{4}D + \frac{1}{2}$	$l = \frac{1}{16}a + \frac{1}{16}$	$r = c + h$
$d = \frac{1}{3}D + \frac{1}{4}$	$m = \frac{11}{16}a - \frac{1}{2}$	$s = c - g$
$e = \frac{1}{64}a + \frac{5}{32}$	$n = \frac{1}{16}a + \frac{1}{16}$	

Problem 1.—Make a drawing showing two views for a 1¼-inch coupling. Show in half-section with all bolts which are visible. Draw full size.

Problem 2.—Make a drawing showing two views for a 3-inch coupling with six bolts. Show in half-section. Scale 6 in. = 1 ft.

Problem 3.—Assume a shaft diameter for a safety coupling and draw two views. Show with the shafts removed.

Inner Gland

30°

Outer Gland

PISTON ROD
STUFFING BOX

FLANGED
SAFETY COUPLING

Cast Iron
NUT COUPLING

Nut couplings may be used on rods whose lengths it may be desirable to vary within certain limits. Couplings of this kind may be made of cast iron, steel, or bronze.

The drawing shows a nut coupling which allows a rod, consisting of two parts, to be lengthened or shortened. One part screws into the coupling with a right-hand thread; the other part screws into the coupling with a left-hand thread.

Study the drawing for method of finding the lines of intersection.

D = diameter of rod	$e = 1\frac{1}{4}D + 1$	$k = D + \frac{5}{8}$
$a = 4D$	$f = 1\frac{1}{4}D + \frac{3}{4}$	$m = \frac{3}{16}a + \frac{1}{4}$
$b = 1\frac{1}{2}D + 1\frac{1}{8}$	$g = \frac{5}{32}a - \frac{3}{8}$	$n = \frac{5}{8}a - \frac{1}{2}$
$c = \frac{1}{4}a + \frac{1}{4}$	$h = \frac{1}{16}a + 2\frac{1}{8}$	$p = \frac{1}{32}D + \frac{9}{64}$
$d = \frac{3}{4}a - \frac{1}{2}$		

Problem 1.—Draw three views of a nut coupling for a $1\frac{3}{8}$-inch rod. Show one view in full-section. Draw full size.

Problem 2.—Draw three views of a nut coupling for a $2\frac{1}{2}$-inch rod. Show one view in half-section. Scale 6 in. = 1 ft.

Problem 3.—Assume a rod diameter for a nut coupling and draw three views. Show a section in one view.

Machine Steel
FORKED COUPLING

Forked rods are sometimes used for the cross-head end of connecting rods for steam engines. Similar rods are also used in a variety of machines. The jaws may be solid, or they may be provided with caps as shown. They are always made of steel.

The jaws on the rod in the drawing are provided with caps which are fastened to the jaws by means of cap screws. (See table for proportions of cap screws.)

Study the drawing for method of finding the various lines of intersection in the lower jaw.

D = diameter of bore	$f = 1\frac{1}{8}e - \frac{21}{32}$	$m = \frac{3}{8}a - \frac{1}{16}$
$a = 2D$	$g = \frac{5}{8}e - 1\frac{1}{2}$	$n = \frac{1}{4}a + \frac{1}{8}$
$b = 5D - 2\frac{3}{8}$	$h = f - g$	$o = \frac{1}{2}a - \frac{7}{16}$
$c = D - \frac{9}{16}$	$i = \frac{1}{2}a + \frac{3}{4}$	$p = \frac{1}{4}a - \frac{7}{16}$
$d = 1\frac{1}{2}D - \frac{5}{8}$	$k = \frac{1}{4}a + 1\frac{1}{4}$	$q = \frac{1}{2}a - \frac{1}{8}$
$e = 2D - \frac{1}{4}$	$l = \frac{1}{2}a - \frac{1}{8}$	$r = n - \frac{7}{16}$
		$s = r + \frac{1}{32}$

Problem 1.—Make a drawing showing three views of a rod for a $1\frac{1}{4}$-inch bore. Draw full size.

Problem 2.—Draw three views of a rod for a $2\frac{3}{4}$-inch bore. Scale 6 in. = 1 ft.

Problem 3.—Assume a bore for a forked rod and draw three views.

PLATE 101

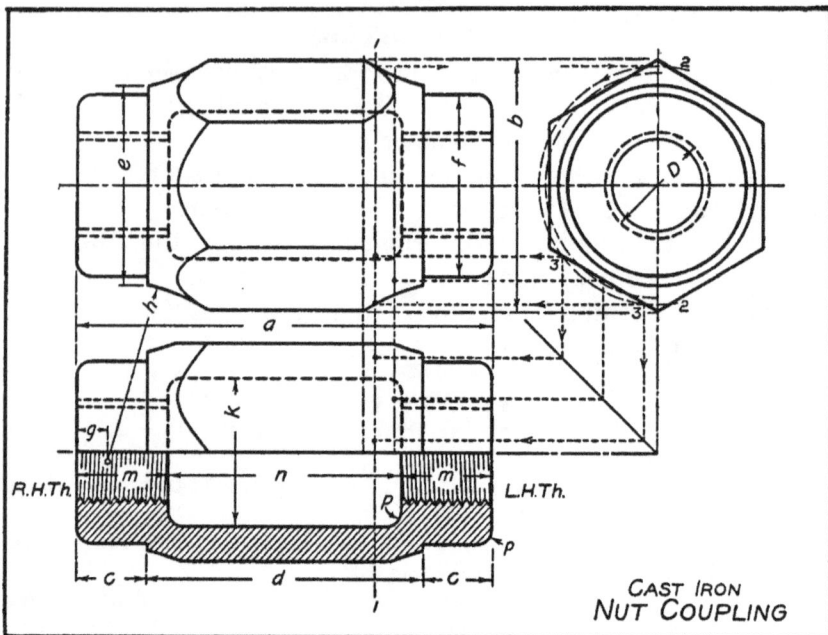

CAST IRON
NUT COUPLING

R.H.Th.   L.H.Th.

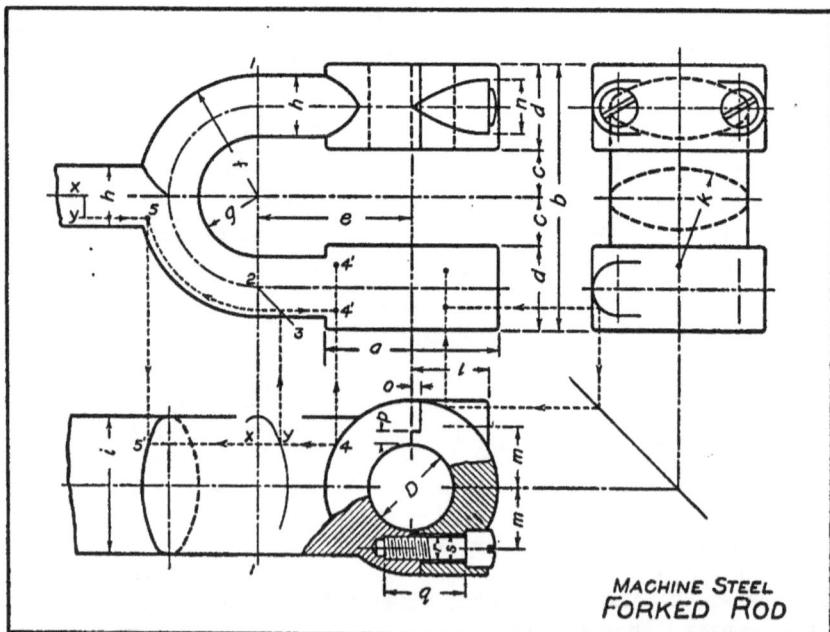

MACHINE STEEL
FORKED ROD

PLATE 102

Clutch Coupling
SHIFTING GEAR

Shifting gears are employed in engaging or disengaging couplings and clutches, and for moving belts from tight or loose pulleys or vice versa.

For D see Clutch Couplings, page 122.

D = diameter of shaft	$h = \frac{1}{4}D + \frac{3}{8}$	$p = \frac{1}{2}D + \frac{3}{8}$
$a = 1\frac{3}{4}D + 1$	$i = \frac{1}{8}D + \frac{1}{4}$	$r = d + \frac{1}{8}$
$b = \frac{1}{2}D + \frac{9}{16}$	$j = \frac{3}{16}D + \frac{1}{4}$	$s = 1\frac{1}{4}d + \frac{3}{16}$
$c = 1\frac{5}{8}D + \frac{7}{8}$	$k = \frac{5}{16}D + \frac{7}{16}$	$t = j + \frac{1}{32}$
$d = \frac{1}{4}D + \frac{3}{16}$	$l = \frac{1}{4}D + \frac{3}{8}$	$u = 2d - \frac{1}{8}$
$e = 1\frac{1}{2}D + \frac{1}{2}$	$m = \frac{3}{8}D + \frac{5}{8}$	$v = \frac{1}{8}D + \frac{3}{8}$
$f = 2D + 1\frac{1}{8}$	$n = \frac{1}{4}D + \frac{3}{16}$	$w = \frac{1}{8}D + \frac{1}{16}$
$g = 1\frac{1}{2}D + 1$	$o = \frac{3}{16}D + \frac{3}{16}$	$x = \frac{1}{8}D + \frac{1}{8}$
		y to be assumed

Problem 1.—Make a drawing as shown. Let $D = 1$ and $y = 5$ inches. Draw full size.

Problem 2.—Make a drawing as shown having the lower half of the collar in section. Let $D = 1\frac{3}{4}$ and $y = 8\frac{3}{4}$ inches. Scale 6 in. = 1 ft.

Problem 3.—Make a drawing as shown. Assume dimensions for D and y; also assume a suitable section.

Buttress-Thread
PLANER JACK

Planer jacks are frequently used on the beds, or tables, of planing machines, milling machines or boring machines, to assist in holding or supporting machine parts which are to be planed, milled or bored.   They are made with a cast iron base or nut, and a soft steel screw and cap.

The drawing shows the top view, the front view, and the center line for the side view of a jack with a buttress-thread screw. Square-thread screws are also frequently used.

D = diameter of screw	$h = \frac{3}{4}D + \frac{1}{16}$	$p = \frac{1}{4}D$
$a = 3\frac{1}{4}D + \frac{1}{16}$	$j = \frac{3}{8}D + \frac{1}{32}$	$r = \frac{1}{2}D + \frac{1}{8}$
$b = 2\frac{1}{4}D + \frac{3}{16}$	$k = \frac{1}{2}D$	$s = \frac{1}{8}D - \frac{1}{32}$
$c = 1\frac{1}{2}D + \frac{1}{8}$	$l = 3\frac{3}{8}D + \frac{5}{32}$	$t = \frac{5}{8}D + \frac{5}{32}$
$e = 1\frac{1}{8}D - \frac{1}{32}$	$m = \frac{1}{4}D + \frac{1}{8}$	$u = \frac{1}{20}D$
$f = \frac{1}{4}D + \frac{1}{16}$	$n = D + \frac{1}{16}$	$v = \frac{1}{2}D$
$g = D$	$o = \frac{1}{8}D + \frac{3}{32}$	

Problem 1.—Draw three views and details of a planer jack with buttress-thread screw of $\frac{1}{2}$-inch pitch. Let $D = \frac{3}{4}$ inch. Draw full size.

Problem 2.—Draw three views and details of a planer jack with a square-thread screw of $\frac{1}{4}$-inch pitch. Let $D = 1$ inch. Scale 6 in. = 1 ft.

Problem 3.—Make a drawing showing three views and necessary details for a planer jack having a square-thread screw. Assume a pitch and a diameter for the screw.

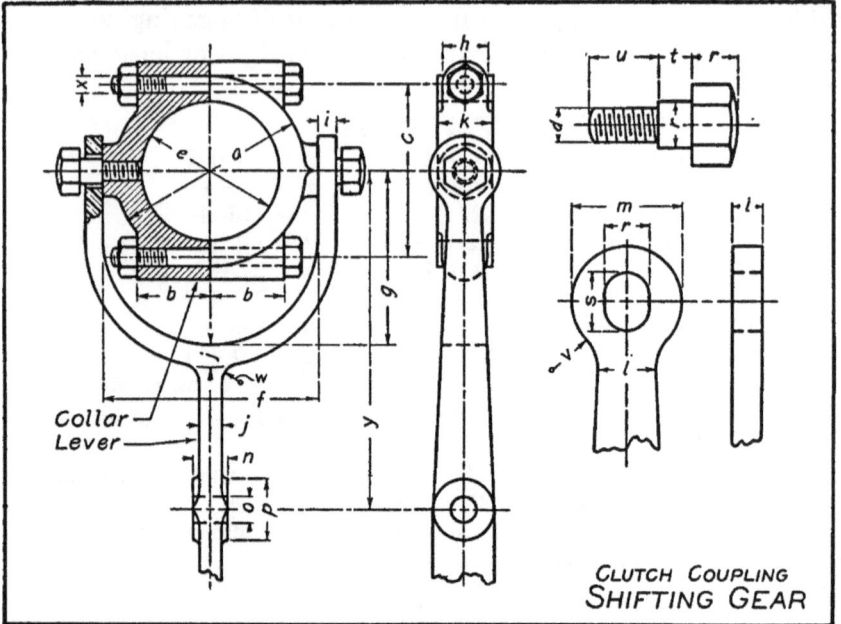

Collar
Lever

CLUTCH COUPLING
SHIFTING GEAR

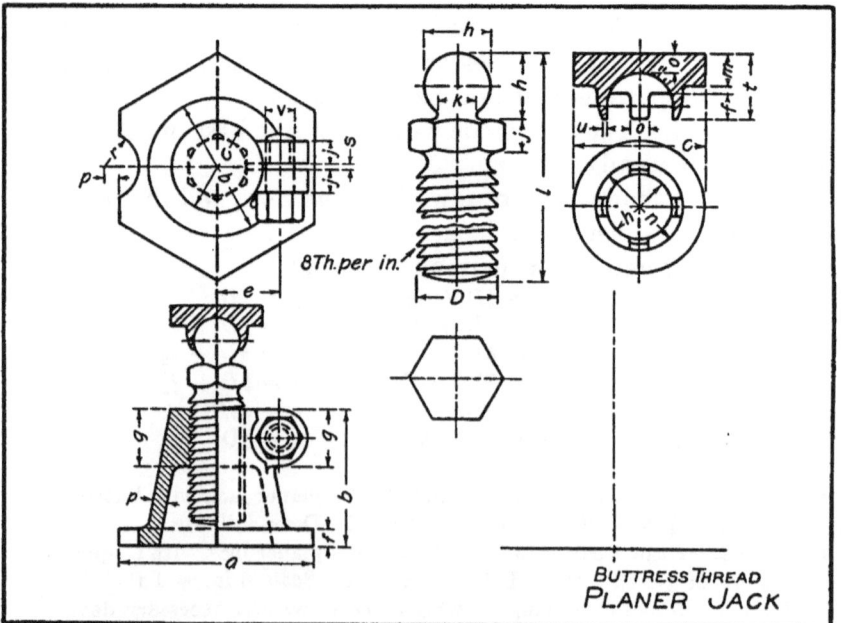

8Th.per in.

BUTTRESS THREAD
PLANER JACK

PLATE 104

Single-Curve Arm
BELT PULLEY

To find the center for one curve of an arm as shown proceed as follows: From x and z draw 30° lines intersecting at 1, and from 1 as center draw arc xyz. Lay off d and d', giving points 2–3 and 4–5. Draw line 2–3. From 3 draw a line making angle 6–3–2 equal angle 1–2–3, giving point 7, the center for curve 2–3. The center for curve 4–5 is found similarly.

A = diameter of pulley	$a = 2C$	$d' = \frac{2}{3}d$
B = width of face	$b = \frac{3}{4}B$	$e = \frac{d'}{4}$
C = diameter of bore	$c = .005B + .03$	$f = \frac{A}{200} + \frac{1}{8}$
N = number of arms	$d = .63\sqrt[3]{\dfrac{AB}{N}}$	$g = \frac{3}{8}$ inch tap

Problem 1.—Make a drawing of a pulley with five arms. Let $A = 9$, $B = 2\frac{1}{4}$ and $C = 1\frac{1}{8}$ inches. Draw full size.

Problem 2.—Make a drawing of a pulley with six arms. Let $A = 16$, $B = 4\frac{1}{2}$ and $C = 2$ inches. Scale 6 in. = 1 ft.

Problem 3.—Make a drawing of a 20-inch pulley having a 5-inch face, 2½-inch bore, and six straight arms.

Double-Curve Arm
BELT PULLEY

To find the centers for the arcs of the lower part of an arm as shown proceed as follows: Draw 45° lines through x, y, z, and draw arcs xy and yz. Lay off d and d'', giving points 1–2 and 3–4. From 3, on line mn, lay off a distance equal to 1–5, giving point 6, the center for curve 1–3. From 4, on line mn, lay off a distance equal to 2–5, giving point 7, the center for curve 2–4. The centers for the upper part of the arm are found similarly.

A = diameter of pulley	$a = 2C$	$d'' = \frac{1}{3}d$
B = width of face	$b' = \frac{2}{3}B$	$e = \frac{d'}{4}$
C = diameter of bore	$c = .005B + .03$	$f = \frac{A}{200} + \frac{1}{8}$
N = number of arms	$d = .63\sqrt[3]{\dfrac{AB}{N}}$	$h = \frac{1}{4}$ inch drill
	$d' = \frac{2}{3}d$	$xy = \frac{2}{3}xz$

Problem 1.—Draw two views of a pulley having five arms similar to those shown. Let $A = 9$, $B \doteq 2$ and $C = 1\frac{1}{4}$ inches. Draw full size.

Problem 2.—Draw two views of a pulley having six arms similar to those shown. Let $A = 18$, $B = 4\frac{1}{2}$ and $C = 2$ inches. Scale 6 in. = 1 ft.

Problem 3.—Draw two views of a 36-inch pulley having a 9-inch face, 3-inch bore, and six arms similar to those shown.

PLATE 105

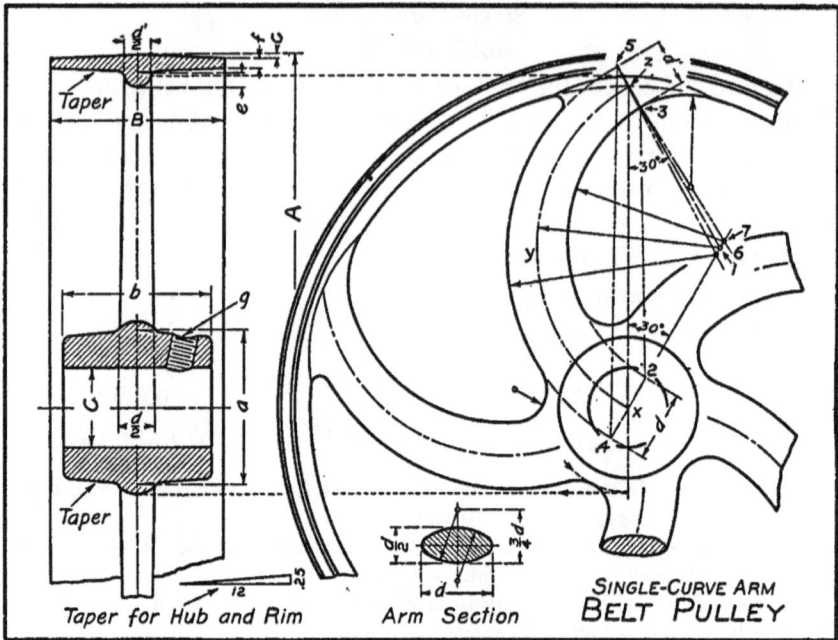

Taper for Hub and Rim

Arm Section

SINGLE-CURVE ARM
BELT PULLEY

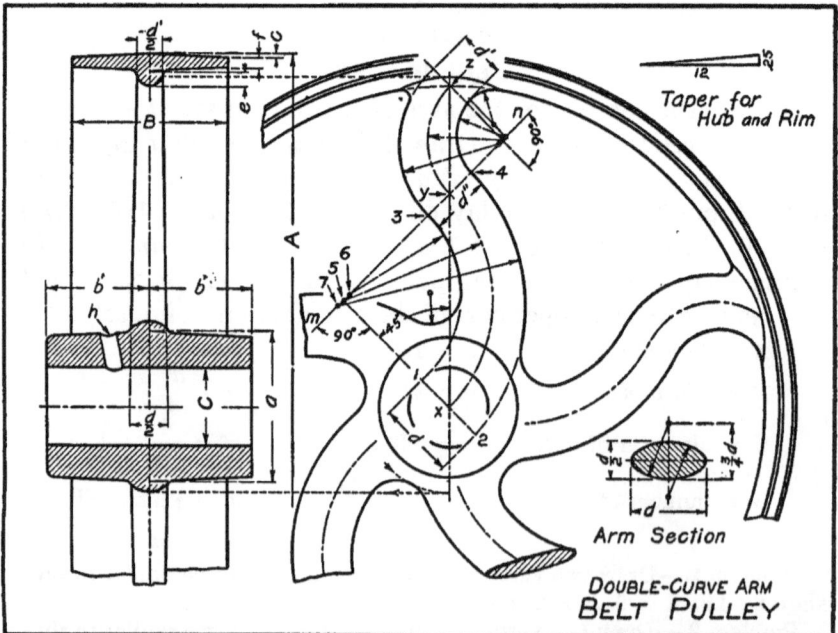

Taper for
Hub and Rim

Arm Section

DOUBLE-CURVE ARM
BELT PULLEY

PLATE 106

Compression
GREASE CUP

Grease cups are receptacles for holding heavy, viscid lubricants. They are placed near or on top of bearings, such as engine bearings, shaft bearings, etc., to provide a means for lubrication. Grease cups are generally made of brass, although cheap grades are made of iron or steel. They are always provided with a pipe-thread for fastening to the bearing.

a = depth of body	$j = \frac{1}{16}a + \frac{1}{8}$	$s = \frac{29}{32}a + \frac{5}{8}$
$b = \frac{2}{3}a + \frac{1}{4}$	$k = \frac{5}{16}a + \frac{5}{16}$	$t = \frac{2}{3}a + \frac{1}{8}$
$c = \frac{27}{32}a + \frac{15}{32}$	$l = 1\frac{1}{4}a + \frac{3}{4}$	$u = \frac{5}{16}a + \frac{3}{16}$
$d = \frac{13}{16}a + \frac{3}{8}$	$m = \frac{1}{2}a + 1\frac{1}{8}$	$v = \frac{1}{2}a + \frac{3}{32}$
$e = 1\frac{1}{3}a + \frac{3}{4}$	$n = \frac{1}{32}a + \frac{1}{16}$	$w = \frac{1}{16}a + \frac{1}{16}$
$f = a + \frac{1}{8}$	$o = \frac{5}{8}a + \frac{1}{16}$	$x = \frac{3}{32}a + \frac{3}{32}$
$g = \frac{5}{8}a + \frac{1}{16}$	$p = \frac{1}{4}a + \frac{1}{4}$	$y = \frac{3}{4}a + \frac{1}{2}$
$h = \frac{1}{2}a + \frac{1}{4}$	$q = \frac{11}{16}a + \frac{1}{16}$	$z = \frac{23}{32}a + \frac{7}{16}$
$i = \frac{1}{16}a + \frac{3}{8}$	$r = \frac{7}{16}a + \frac{1}{8}$	

Problem 1.—Draw a grease cup and details as shown. Let $a = 2\frac{1}{4}$ inches, and $y' = \frac{3}{8}$-inch pipe thread. Draw full size.

Problem 2.—Draw the assembled view of a grease-cup in half-section, and the details as shown. Let $a = 4$ inches, and $y' = \frac{1}{2}$-inch pipe thread. Scale 6 in. = 1 ft.

Problem 3.—Assume a value for a and a diameter for the pipe thread, and make the necessary views for a working drawing of a screw-feed grease cup.

Adjustable
LATHE CHUCK

Lathe chucks are devices which are screwed to the spindles of lathes and are used for holding materials to be turned, bored or drilled. The drawing shows an adjustable chuck suitable for use on a wood-turning lathe.

a = diameter of spindle thread		m = 10 threads per inch	
b = length of spindle thread		n = diameter of drill for No. 12 screw	
c = number of threads per inch			
$d = a + \frac{3}{4}$	$h = \frac{1}{4}$	$q = b + \frac{3}{4}$	$u = \frac{3}{4}$
$e = b + \frac{1}{16}$	$k = a + \frac{1}{16}$	$r = d + \frac{5}{8}$	$v = \frac{1}{4}$
$f = \frac{3}{16}$	$l = b + \frac{3}{4}$	$s = q - \frac{7}{16}$	$w = \frac{5}{16}$
$g = \frac{1}{2}$	$p = d + 1\frac{1}{2}$	$t = \frac{3}{16}$	$x = r + \frac{1}{8}$
			$y = \frac{3}{8}$

Problem 1.—Make an assembled drawing for a chuck, and draw details as shown. Let $a = 1\frac{1}{4}$ inches, $b = 1\frac{3}{4}$ inches, and $c = 8$ threads per inch.

Problem 2.—Make an assembled drawing for a chuck in half-section, and draw details as shown. Let $a = 1\frac{3}{4}$ inches, $b = 1\frac{1}{2}$ inches, and $c = 8$ threads per inch.

Problem 3.—Measure the spindle of a lathe for a, b, and c, and make a working drawing for an adjustable chuck as shown.

PLATE 107

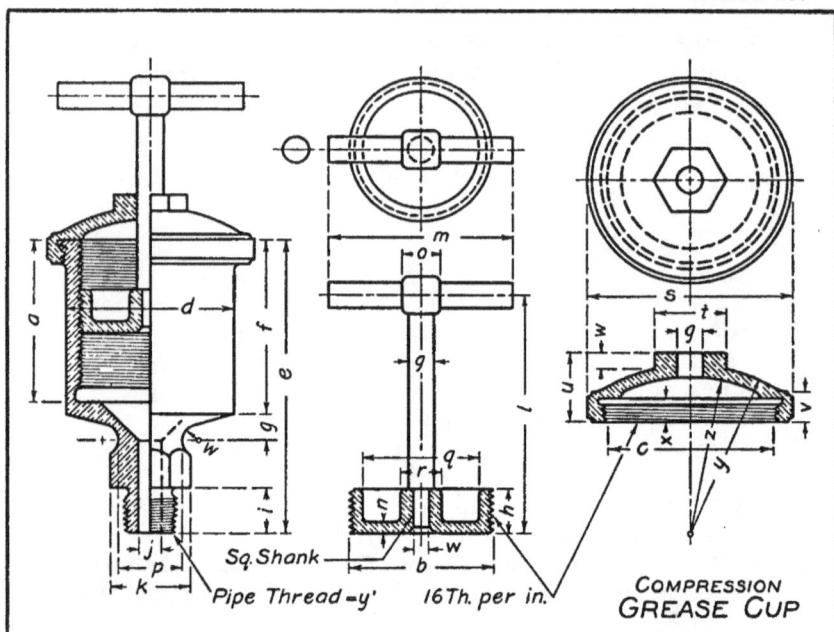

Sq. Shank

Pipe Thread –y'   16 Th. per in.

COMPRESSION
GREASE CUP

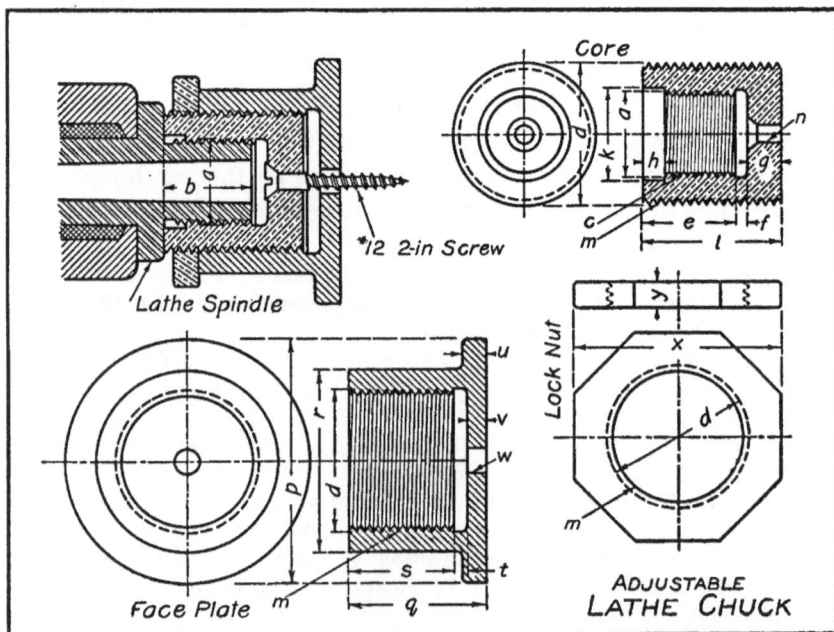

Core

*12 2-in Screw

Lathe Spindle

Lock Nut

Face Plate

ADJUSTABLE
LATHE CHUCK

PLATE 108

PIPE UNIONS

Pipe unions are metallic fittings used to join sections of pipe; as water pipe, gas pipe, steam pipe, etc. They may be classified as nut unions and flange unions. Nut unions are generally used for pipes to 2 inches diameter, and flange unions for pipes above 2 inches, although nut unions for pipes to 4-inch diameter may be obtained. Nut unions are made of malleable iron, malleable iron and brass, and all brass. Flange unions are made of cast iron.

Spherical Seat
PIPE UNION

This drawing shows a nut union with a spherical seat. The seat is made of two brass rings which are accurately ground to insure a tight joint.

D = diameter of pipe

$$a = 1\tfrac{7}{16}D + \tfrac{1}{8}$$
$$b = \tfrac{5}{16}D + \tfrac{11}{16}$$
$$c = \tfrac{7}{16}D + \tfrac{13}{16}$$
$$d = 1\tfrac{5}{16}D + \tfrac{11}{16}$$
$$e = 1\tfrac{3}{16}D + \tfrac{3}{8}$$

$$f = 1\tfrac{3}{16}D + \tfrac{7}{16}$$
$$g = 1\tfrac{3}{16}D + \tfrac{3}{8}$$
$$h = 1\tfrac{7}{32}D + \tfrac{19}{32}$$
$$i = \tfrac{1}{16}D + \tfrac{1}{16}$$
$$j = \tfrac{1}{8}D + \tfrac{1}{8}$$
$$k = \tfrac{1}{8}D + \tfrac{3}{8}$$

$$l = \tfrac{1}{8}D + \tfrac{1}{16}$$
$$m = 1\tfrac{1}{8}D + \tfrac{1}{16}$$
$$n = D + \tfrac{1}{8}$$
$$o = \tfrac{3}{8}D + \tfrac{3}{16}$$
$$p = 1\tfrac{3}{16}D + \tfrac{1}{2}$$
$$r = D + \tfrac{3}{4}$$

Problem 1.—Make a drawing showing an assembly section, and details of a nut union for a 1-inch pipe. Draw full size.

Problem 2.—Make a drawing showing an assembly in half-section, and details in full-section of a nut union for a 2-inch pipe. Scale 6 in. = 1 ft.

Problem 3.—Assume a pipe diameter and make a working drawing for a nut union with a spherical seat.

Gasket Seat
PIPE UNION

This drawing shows a nut union with a gasket seat. Gaskets are generally made of a rubber composition, or of asbestos. They insure a tight, non-leaking joint.

D = diameter of pipe

$$a = 1\tfrac{1}{2}D + \tfrac{15}{16}$$
$$b = \tfrac{1}{4}D + \tfrac{3}{4}$$
$$c = \tfrac{3}{8}D + 1$$
$$d = 1\tfrac{3}{8}D + \tfrac{3}{4}$$

$$e = 1\tfrac{3}{16}D + \tfrac{3}{8}$$
$$f = 1\tfrac{5}{16}D + \tfrac{5}{8}$$
$$g = 1\tfrac{1}{8}D + \tfrac{7}{16}$$
$$h = D$$
$$i = \tfrac{1}{16}D + \tfrac{1}{16}$$

$$j = \tfrac{1}{8}D + \tfrac{1}{4}$$
$$k = \tfrac{1}{8}D + \tfrac{3}{4}$$
$$l = \tfrac{1}{16}D + \tfrac{1}{8}$$
$$m = 1\tfrac{5}{16}D + \tfrac{7}{16}$$
$$n = \tfrac{1}{8}D + \tfrac{1}{2}$$

Problem 1.—Make a drawing showing front, top, and end views of a nut union for a 1-inch pipe. Show front and top views in half-section.

Problem 2.—Make a drawing showing three views of a nut union for a 2-inch pipe. Assume suitable sections. Scale 6 in. = 1 ft.

Problem 3.—Assume a pipe diameter, and make a working drawing for a nut union with a gasket seat.

PLATE 109

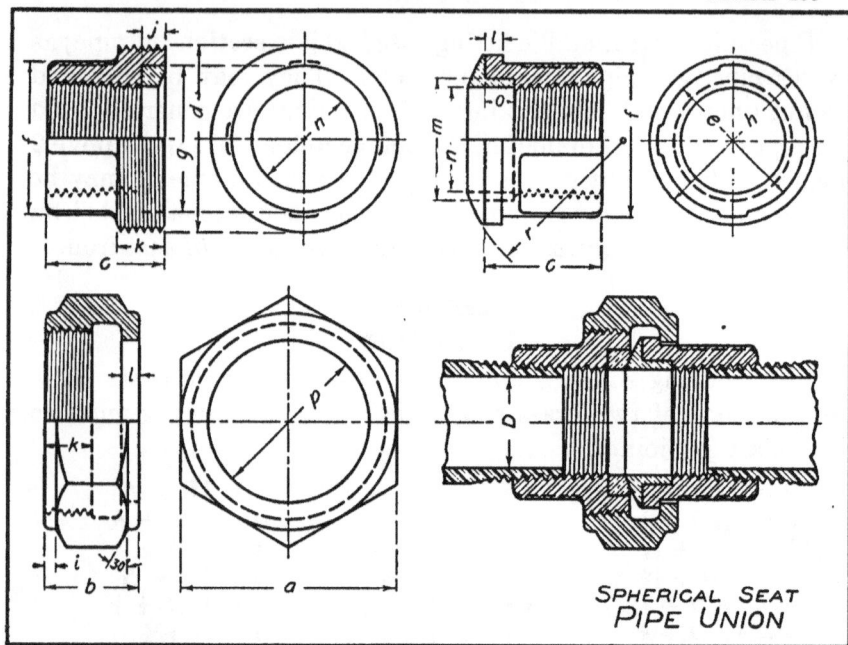

SPHERICAL SEAT
PIPE UNION

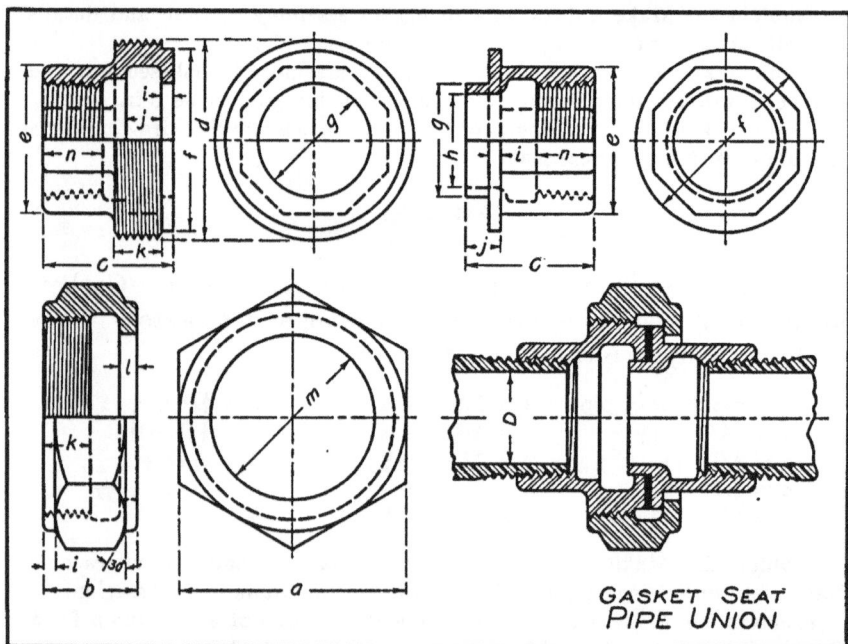

GASKET SEAT
PIPE UNION

PLATE 110

Square Thread
SCREW JACK

Screw jacks are portable devices used for lifting heavy loads through short distances. To raise a load, the screw is turned with a bar inserted into the holes of the screw head.

The drawing shows a simple screw jack. It consists of a square-threaded steel screw working in a cast iron body or nut.

Problem 1.—Make an assembly drawing with details for a jack as shown. Let $a = 4\frac{1}{2}$ inches and $b = 6$ inches. Draw full size.

Problem 2.—Make an assembly drawing as shown, and draw details of the screw, bearing plate, and bearing plate screw. Let $a = 6$ inches and $b = 7\frac{1}{2}$ inches. Draw full size.

Problem 3.—Multiply the dimensions shown by 1.5 and make an assembly drawing for a jack with a buttress-thread screw of one-eighth pitch, and draw details of the screw, bearing plate, and bearing plate screw. Change calculated dimensions to nearest $\frac{1}{16}$ or $\frac{1}{8}$ inch, according to judgment. Let $a = 7$ inches and $b = 8\frac{3}{4}$ inches. Scale 6 in. = 1 ft.

BALL BEARING

Ball bearings are used in many machines to reduce frictional resistance between bearings and journals, or shafts.

The drawing shows three views and details of a ball bearing suitable for a horizontal one-inch shaft. The races are of hardened steel and the casing is of cast iron. The inner race is securely fastened by means of a nut threaded on the end of the shaft, while the outer races are secured, after being properly adjusted, by tightening the casing with a stud and nut as shown.

Problem 1.—Make an assembly drawing having three views and details of a bearing as shown. Let $a = 2\frac{3}{4}$ inches.

Problem 2.—Draw front and top views of a bearing with the dimensions shown. Show top view in full-section at the axis of the shaft. Draw two views of the inner race and two views of one outer race showing each in half-section. Let $a = 3$ inches.

Problem 3.—Make a detail drawing of the bearing shown. Show three views of the casing, two views of the inner race, two views of one outer race, and one view of the stud and nut. Let $a = 2\frac{3}{4}$ inches.

PLATE 111

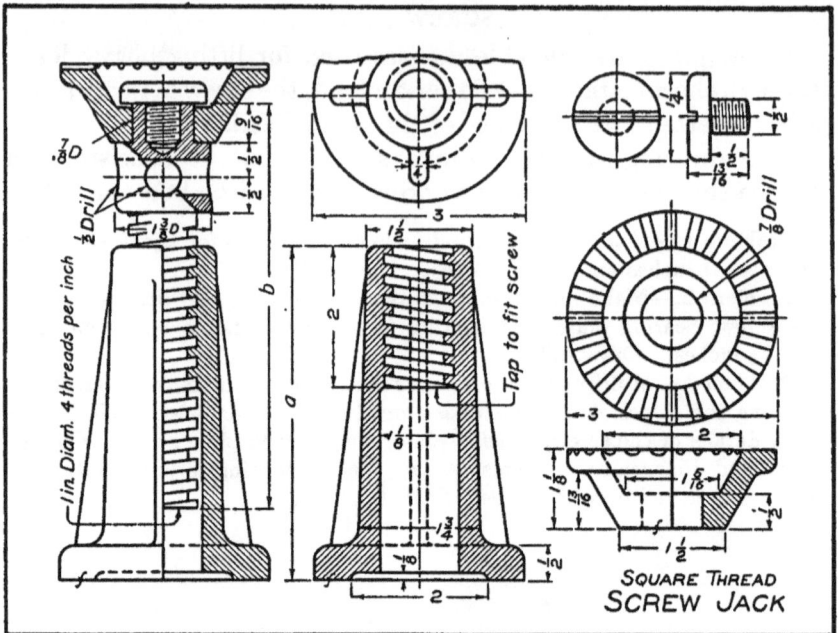

SQUARE THREAD
SCREW JACK

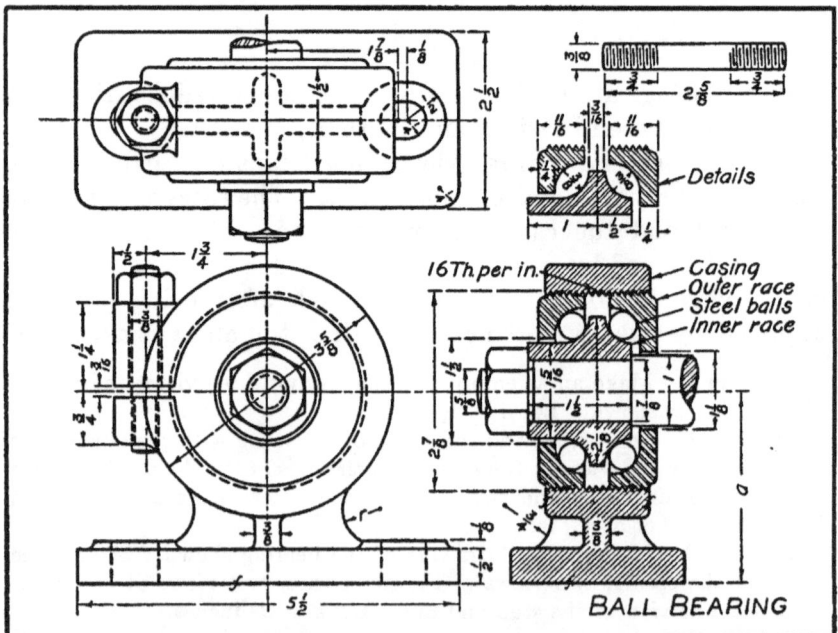

BALL BEARING

PLATE 112

PART III
TABLES

List of Tables

TABLE 1. CAP SCREWS

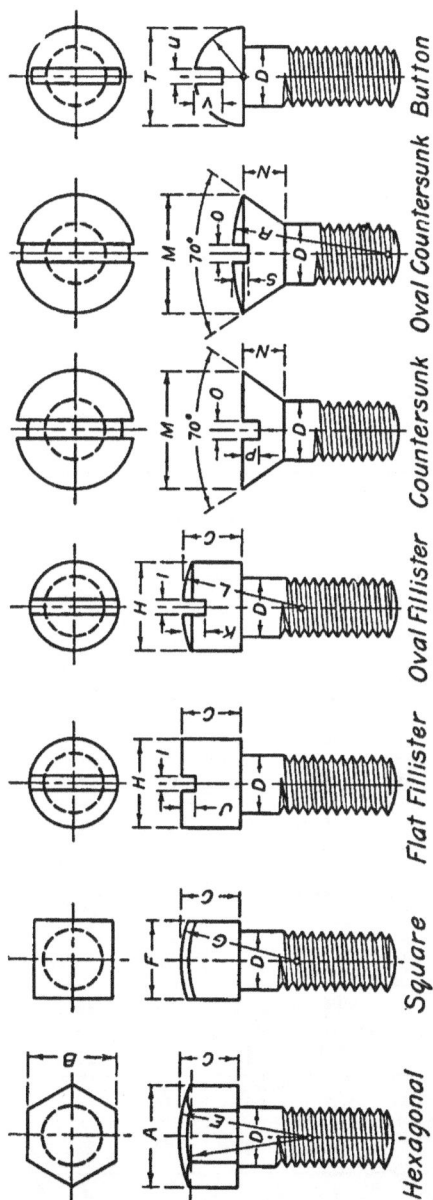

Hexagonal   Square   Flat Fillister   Oval Fillister   Countersunk   Oval Countersunk   Button

n = number of threads per inch   d = diameter of tap drill.

D	n	d	A	B	C	E	F	G	H	I	J	K	L	M	N	O	P	R	S	T	U	V
	40									.032						.040					.035	
	32									.040						.064					.051	
	20									.064						.072					.072	
	18									.072						.102					.091	
	16									.091						.114					.102	
	14									.102						.114					.114	
	13									.114						.128					.114	
	12									.114						.133					.114	
	11									.128						.133					.133	
	10									.133						.133					.133	

TABLE II.—U. S. STANDARD BOLTS AND NUTS

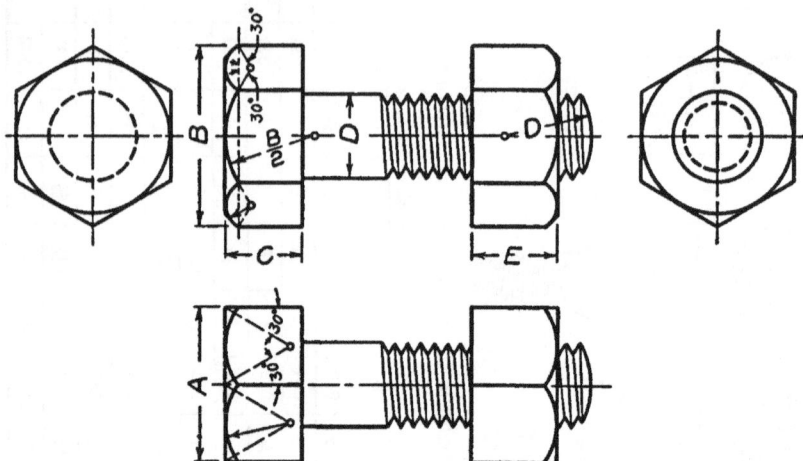

$$A = 1\tfrac{1}{2}\,D + \tfrac{1}{8} \qquad B = 1\tfrac{3}{4}\,D + \tfrac{1}{8}$$

$$C = \frac{A}{2} \qquad E = D$$

D = Diameter	No. of threads per inch	Area at root of thread, sq. in.	Rough A	Rough B	Rough C	Rough E	Finished A	Finished B	Finished C	Finished E
$\frac{1}{4}$	20	.026	$\frac{1}{2}$	$\frac{37}{64}$	$\frac{1}{4}$	$\frac{1}{4}$	$\frac{7}{16}$	$\frac{1}{2}$	$\frac{3}{16}$	$\frac{3}{16}$
$\frac{5}{16}$	18	.045	$\frac{19}{32}$	$\frac{11}{16}$	$\frac{19}{64}$	$\frac{5}{16}$	$\frac{17}{32}$	$\frac{39}{64}$	$\frac{1}{4}$	$\frac{1}{4}$
$\frac{3}{8}$	16	.068	$\frac{11}{16}$	$\frac{51}{64}$	$\frac{11}{32}$	$\frac{3}{8}$	$\frac{5}{8}$	$\frac{23}{32}$	$\frac{5}{16}$	$\frac{5}{16}$
$\frac{7}{16}$	14	.093	$\frac{25}{32}$	$\frac{29}{32}$	$\frac{25}{64}$	$\frac{7}{16}$	$\frac{23}{32}$	$\frac{53}{64}$	$\frac{3}{8}$	$\frac{3}{8}$
$\frac{1}{2}$	13	.126	$\frac{7}{8}$	$1\frac{1}{64}$	$\frac{7}{16}$	$\frac{1}{2}$	$\frac{13}{16}$	$\frac{15}{16}$	$\frac{7}{16}$	$\frac{7}{16}$
$\frac{9}{16}$	12	.162	$\frac{31}{32}$	$1\frac{1}{8}$	$\frac{31}{64}$	$\frac{9}{16}$	$\frac{29}{32}$	$1\frac{1}{64}$	$\frac{1}{2}$	$\frac{1}{2}$
$\frac{5}{8}$	11	.202	$1\frac{1}{16}$	$1\frac{15}{64}$	$\frac{17}{32}$	$\frac{5}{8}$	1	$1\frac{5}{32}$	$\frac{9}{16}$	$\frac{9}{16}$
$\frac{3}{4}$	10	.302	$1\frac{1}{4}$	$1\frac{27}{64}$	$\frac{5}{8}$	$\frac{3}{4}$	$1\frac{3}{16}$	$1\frac{21}{64}$	$\frac{11}{16}$	$\frac{11}{16}$
$\frac{7}{8}$	9	.419	$1\frac{7}{16}$	$1\frac{43}{64}$	$\frac{23}{32}$	$\frac{7}{8}$	$1\frac{3}{8}$	$1\frac{19}{32}$	$\frac{13}{16}$	$\frac{13}{16}$
1	8	.551	$1\frac{5}{8}$	$1\frac{7}{8}$	$\frac{13}{16}$	1	$1\frac{9}{16}$	$1\frac{13}{16}$	$\frac{15}{16}$	$\frac{15}{16}$
$1\frac{1}{8}$	7	.693	$1\frac{13}{16}$	$2\frac{3}{32}$	$\frac{29}{32}$	$1\frac{1}{8}$	$1\frac{5}{8}$	$2\frac{1}{64}$	$1\frac{1}{16}$	$1\frac{1}{16}$
$1\frac{1}{4}$	7	.889	2	$2\frac{5}{16}$	1	$1\frac{1}{4}$	$1\frac{15}{16}$	$2\frac{15}{64}$	$1\frac{3}{16}$	$1\frac{3}{16}$

TABLE III.—MACHINE SCREWS

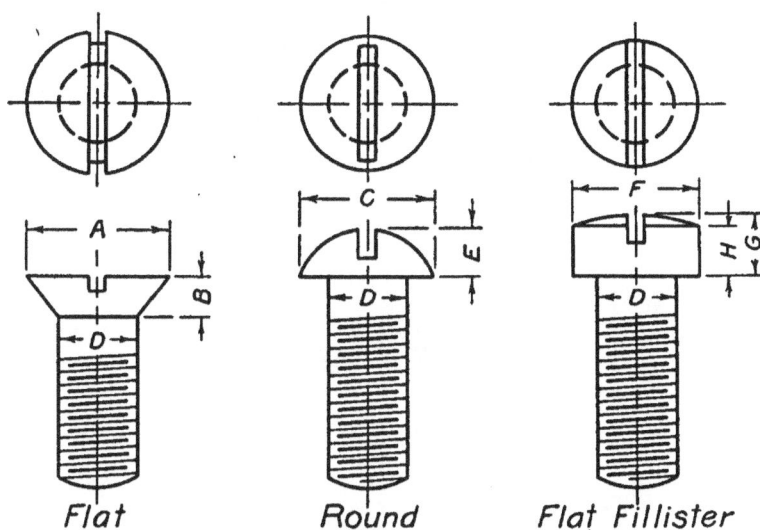

Flat   Round   Flat Fillister

A.S.M.E. STANDARD

No.	No. of Threads	D	A	B	C	E	F	G	H
0	80	.060	.112	.029	.106	.042	.0894	.0496	.0376
1	72	.073	.138	.037	.130	.051	.1107	.0609	.0461
2	64	.086	.164	.045	.154	.060	.132	.0725	.0548
3	56	.099	.190	.052	.178	.069	.153	.0838	.0633
4	48	.112	.216	.060	.202	.078	.1747	.0953	.0719
5	44	.125	.242	.067	.226	.087	.196	.1068	.0805
6	40	.138	.262	.075	.250	.096	.217	.1180	.089
7	36	.151	.294	.082	.274	.105	.2386	.1296	.0976
8	36	.164	.320	.090	.298	.114	.2599	.1410	.1062
9	32	.177	.346	.097	.322	.123	.2813	.1524	.1148
10	30	.190	.372	.105	.346	.133	.3026	.1639	.1234
12	28	.216	.424	.120	.394	.151	.3452	.1868	.1405
14	24	.242	.472	.135	.443	.169	.3879	.2097	.1577
16	22	.268	.528	.150	.491	.187	.4305	.2325	.1748
18	20	.294	.580	.164	.539	.205	.4731	.2554	.192
20	20	.320	.632	.179	.587	.224	.5158	.2783	.2092
22	18	.346	.682	.194	.635	.242	.5584	.3011	.2263
24	16	.372	.732	.209	.683	.260	.601	.3240	.2435
26	16	.398	.788	.224	.731	.278	.6437	.3469	.2606
28	14	.424	.840	.239	.779	.296	.6863	.3698	.2778
30	14	.450	.892	.254	.827	.315	.727	.4024	.295

TABLE IV.—BRIGGS STANDARD PIPE THREADS

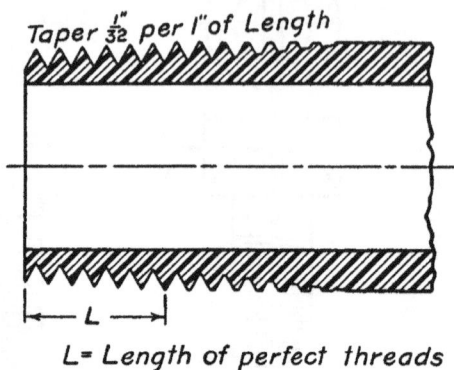

Taper $\frac{1''}{32}$ per 1" of Length

L= Length of perfect threads

Nominal inside diam.	Actual inside diam.	Actual outside diam.	No. of threads per inch	Internal area	Length, perfect threads	Diam. of drill
$\frac{1}{8}$	.270	.405	27	057	$\frac{3}{16}$	$\frac{21}{44}$
$\frac{1}{4}$	.364	.540	18	.104	$\frac{9}{32}$	$\frac{29}{44}$
$\frac{3}{8}$	.494	.675	18	.191	$\frac{19}{64}$	$\frac{19}{44}$
$\frac{1}{2}$	.623	.840	14	.304	$\frac{3}{8}$	$\frac{23}{32}$
$\frac{3}{4}$	.824	1.050	14	.533	$\frac{13}{32}$	$\frac{15}{16}$
1	1.048	1.315	$11\frac{1}{2}$	.861	$\frac{1}{2}$	$1\frac{3}{16}$
$1\frac{1}{4}$	1.380	1.660	$11\frac{1}{2}$	1.496	$\frac{35}{64}$	$1\frac{1}{32}$
$1\frac{1}{2}$	1.610	1.900	$11\frac{1}{2}$	2.036	$\frac{9}{16}$	$1\frac{23}{32}$
2	2.067	2.375	$11\frac{1}{2}$	3.356	$\frac{37}{64}$	$2\frac{3}{16}$
$2\frac{1}{2}$	2.468	2.875	8	4.780	$\frac{57}{64}$	$2\frac{11}{16}$
3	3.067	3.500	8	7.383	$\frac{51}{64}$	$3\frac{5}{16}$
$3\frac{1}{2}$	3.548	4.000	8	9.887	1	$3\frac{13}{16}$
4	4.026	4.500	8	12.730	$1\frac{1}{16}$	$4\frac{3}{16}$
$4\frac{1}{2}$	4.508	5.000	8	15.961	$1\frac{7}{32}$	$4\frac{11}{16}$
5	5.045	5.563	8	19.986	$1\frac{5}{32}$	$5\frac{1}{4}$

TABLE V.—SET SCREWS

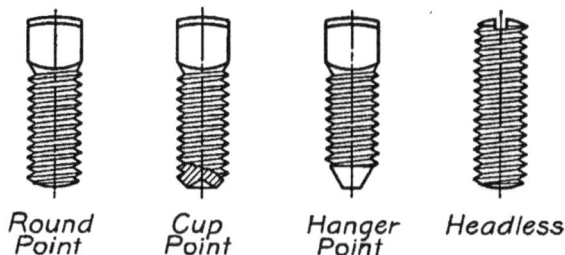

Round Point   Cup Point   Hanger Point   Headless

TABLE VI.—GIB KEYS

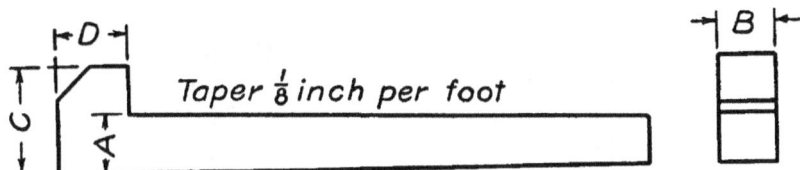

Taper $\frac{1}{8}$ inch per foot

A	B	C	D	A	B	C	D	A	B	C	D
$\frac{1}{8}$	$\frac{1}{8}$	$\frac{1}{4}$	$\frac{7}{32}$	$\frac{7}{16}$	$\frac{7}{16}$	$\frac{3}{4}$	$\frac{17}{32}$	$\frac{3}{4}$	$\frac{3}{4}$	$1\frac{1}{4}$	$\frac{7}{8}$
$\frac{3}{16}$	$\frac{3}{16}$	$\frac{5}{16}$	$\frac{9}{32}$	$\frac{1}{2}$	$\frac{1}{2}$	$\frac{7}{8}$	$\frac{19}{32}$	$\frac{13}{16}$	$\frac{13}{16}$	$1\frac{5}{16}$	$\frac{15}{16}$
$\frac{1}{4}$	$\frac{1}{4}$	$\frac{11}{32}$	$\frac{11}{32}$	$\frac{9}{16}$	$\frac{9}{16}$	1	$\frac{21}{32}$	$\frac{7}{8}$	$\frac{7}{8}$	$1\frac{1}{2}$	1
$\frac{5}{16}$	$\frac{5}{16}$	$\frac{9}{16}$	$\frac{13}{32}$	$\frac{5}{8}$	$\frac{5}{8}$	$1\frac{1}{8}$	$\frac{23}{32}$	$\frac{15}{16}$	$\frac{15}{16}$	$1\frac{5}{8}$	$1\frac{1}{16}$
$\frac{3}{8}$	$\frac{3}{8}$	$\frac{11}{16}$	$\frac{15}{32}$	$\frac{11}{16}$	$\frac{11}{16}$	$1\frac{3}{16}$	$\frac{25}{32}$	1	1	$1\frac{3}{4}$	$1\frac{1}{8}$

TABLE VII.—FEATHER KEYS OR SPLINES

Diameter of shaft.....	1	$1\frac{1}{4}$	$1\frac{1}{2}$	$1\frac{3}{4}$	2	$2\frac{1}{2}$	3	$3\frac{1}{2}$	4	5	6
Width of feather......	$\frac{1}{4}$	$\frac{5}{16}$	$\frac{3}{8}$	$\frac{7}{16}$	$\frac{1}{2}$	$\frac{5}{8}$	$\frac{3}{4}$	$\frac{7}{8}$	1	$1\frac{1}{4}$	$6\frac{3}{8}$
Thickness of feather .	$\frac{3}{8}$	$\frac{7}{16}$	$\frac{1}{2}$	$\frac{9}{16}$	$\frac{5}{8}$	$\frac{3}{4}$	$\frac{7}{8}$	1	$1\frac{1}{4}$	$1\frac{3}{8}$	$1\frac{3}{8}$

TABLE VIII.—AUTOMOBILE SCREWS AND NUTS

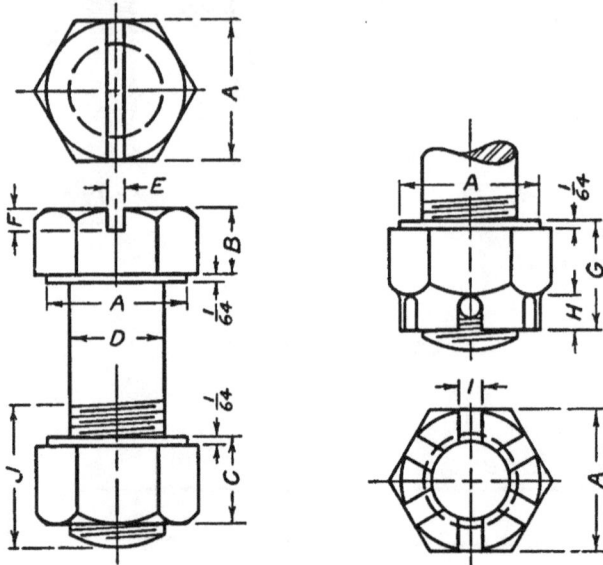

S. A. E. Standard

n = no. of threads per inch;   d = diam. cotter pin

D	A	B	C	E	F	G	H	I	J	n	d
$\frac{1}{4}$	$\frac{7}{16}$	$\frac{3}{16}$	$\frac{3}{32}$	$\frac{1}{16}$	$\frac{3}{32}$	$\frac{9}{32}$	$\frac{3}{32}$	$\frac{5}{64}$	$\frac{3}{8}$	28	$\frac{1}{16}$
$\frac{5}{16}$	$\frac{1}{2}$	$\frac{11}{64}$	$\frac{17}{64}$	$\frac{1}{16}$	$\frac{7}{64}$	$\frac{21}{64}$	$\frac{3}{32}$	$\frac{5}{64}$	$\frac{13}{32}$	24	$\frac{1}{16}$
$\frac{3}{8}$	$\frac{9}{16}$	$\frac{9}{32}$	$\frac{21}{64}$	$\frac{3}{32}$	$\frac{1}{8}$	$\frac{11}{32}$	$\frac{1}{8}$	$\frac{1}{8}$	$\frac{9}{16}$	24	$\frac{3}{32}$
$\frac{7}{16}$	$\frac{5}{8}$	$\frac{21}{64}$	$\frac{3}{8}$	$\frac{3}{32}$	$\frac{1}{8}$	$\frac{27}{64}$	$\frac{1}{8}$	$\frac{1}{8}$	$\frac{21}{32}$	20	$\frac{3}{32}$
$\frac{1}{2}$	$\frac{3}{4}$	$\frac{3}{8}$	$\frac{7}{16}$	$\frac{3}{32}$	$\frac{1}{8}$	$\frac{9}{16}$	$\frac{3}{16}$	$\frac{1}{8}$	$\frac{3}{4}$	20	$\frac{3}{32}$
$\frac{9}{16}$	$\frac{7}{8}$	$\frac{27}{64}$	$\frac{31}{64}$	$\frac{3}{32}$	$\frac{1}{8}$	$\frac{39}{64}$	$\frac{3}{16}$	$\frac{5}{32}$	$\frac{27}{32}$	18	$\frac{1}{8}$
$\frac{5}{8}$	$\frac{15}{16}$	$\frac{15}{32}$	$\frac{35}{64}$	$\frac{3}{32}$	$\frac{1}{8}$	$\frac{21}{32}$	$\frac{1}{4}$	$\frac{5}{32}$	$\frac{15}{16}$	18	$\frac{1}{8}$
$\frac{11}{16}$	1	$\frac{33}{64}$	$\frac{19}{32}$	$\frac{3}{32}$	$\frac{1}{8}$	$\frac{45}{64}$	$\frac{1}{4}$	$\frac{5}{32}$	$1\frac{1}{32}$	16	$\frac{1}{8}$
$\frac{3}{4}$	$1\frac{1}{16}$	$\frac{9}{16}$	$\frac{21}{32}$	$\frac{3}{32}$	$\frac{1}{8}$	$\frac{13}{16}$	$\frac{1}{4}$	$\frac{5}{32}$	$1\frac{1}{8}$	16	$\frac{1}{8}$
$\frac{7}{8}$	$1\frac{1}{4}$	$\frac{21}{32}$	$\frac{49}{64}$	$\frac{3}{32}$	$\frac{1}{8}$	$\frac{29}{32}$	$\frac{1}{4}$	$\frac{5}{32}$	$1\frac{5}{16}$	14	$\frac{1}{8}$
1	$1\frac{7}{16}$	$\frac{3}{4}$	$\frac{7}{8}$	$\frac{3}{32}$	$\frac{1}{8}$	1	$\frac{1}{4}$	$\frac{5}{32}$	$1\frac{1}{2}$	14	$\frac{1}{8}$

TABLE IX.—JARNO TAPERS

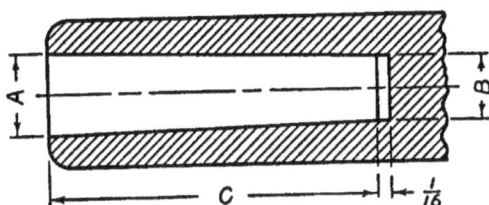

No.	A	B	C	No.	A	B	C	No.	A	B	C
2	.250	.20	1	9	1.125	.90	4½	16	2.000	1.60	8
3	.375	.30	1½	10	1.250	1.00	5	17	2.125	1.70	8½
4	.500	.40	2	11	1.375	1.10	5½	18	2.250	1.80	9
5	.625	.50	2½	12	1.500	1.20	6	19	2.375	1.90	9½
6	.750	.60	3	13	1.625	1.30	6½	20	2.500	2.00	10
7	.875	.70	3½	14	1.750	1.40	7				
8	1.000	.80	4	15	1.875	1.50	7½				

TABLE X.—MORSE TAPERS

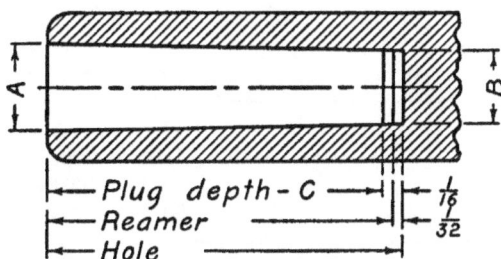

No.	A	B	C	No.	A	B	C	No.	A	B	C
0	.356	.252	2	3	.938	.778	$3\frac{3}{16}$	6	2.494	2.116	7¼
1	.475	.369	2⅛	4	1.231	1.020	$4\frac{1}{16}$	7	3.270	2.750	10
2	.700	.572	$2\frac{9}{16}$	5	1.748	1.475	$5\frac{3}{16}$				

TABLE XI.—DECIMALS OF AN INCH FOR EACH $\frac{1}{64}$TH

$\frac{1}{32}$ds.	$\frac{1}{64}$ths.	Decimal.	Fraction.	$\frac{1}{32}$ds.	$\frac{1}{64}$ths.	Decimal.	Fraction.
	1	.015625			33	.515625	
1	2	.03125		17	34	.53125	
	3	.046875			35	.546875	
2	4	.0625	1–16	18	36	.5625	9–16
	5	.078125			37	.578125	
3	6	.09375		19	38	.59375	
	7	.109375			39	.609375	
4	8	.125	1–8	20	40	.625	5–8
	9	.140625			41	.640625	
5	10	.15625		21	42	.65625	
	11	.171875			43	.671875	
6	12	.1875	3–16	22	44	.6875	11–16
	13	.203125			45	.703125	
7	14	.21875		23	46	.71875	
	15	.234375			47	.734375	
8	16	.25	1–4	24	48	.75	3–4
	17	.265625			49	.765625	
9	18	.28125		25	50	.78125	
	19	.296875			51	.796875	
10	20	.3125	5–16	23	52	.8125	13–16
	21	.328125			53	.828125	
11	22	.34375		27	54	.84375	
	23	.359375			55	.859375	
12	24	.375	3–8	28	56	.875	7–8
	25	.390625			57	.890625	
13	26	.40625		29	58	.90625	
	27	.421875			59	.921875	
14	28	.4375	7–16	30	60	.9375	15–16
	29	.453125			61	.953125	
15	30	.46875		31	62	.96875	
	31	.484375			63	.984375	
16	32	.5	1–2	32	64	1.	

TABLE XII.—AREAS AND CIRCUMFERENCES OF CIRCLES FROM 1 TO 10

Dia.	Area	Circum.	Dia.	Area	Circum.	Dia.	Area	Circum.
1/32	0.00077	0.098175	2	3.1416	6.28319	5	19.635	15.7080
3/64	0.00173	0.147262	1/16	3.3410	6.47953	1/16	20.129	15.9043
1/16	0.00307	0.196350	1/8	3.5466	6.67588	1/8	20.629	16.1007
3/32	0.00690	0.294524	3/16	3.7583	6.87223	3/16	21.135	16.2970
1/8	0.01227	0.392699	1/4	3.9761	7.06858	1/4	21.648	16.4934
5/32	0.01917	0.490874	5/16	4.2000	7.26493	5/16	22.166	16.6897
3/16	0.02761	0.589049	3/8	4.4301	7.46128	3/8	22.691	16.8861
7/32	0.03758	0.687223	7/16	4.6664	7.65763	7/16	23.221	17.0824
1/4	0.04909	0.785398	1/2	4.9087	7.85398	1/2	23.758	17.2788
9/32	0.06213	0.883573	9/16	5.1572	8.05033	9/16	24.301	17.4751
5/16	0.07670	0.981748	5/8	5.4119	8.24668	5/8	24.850	17.6715
11/32	0.09281	1.07992	11/16	5.6727	8.44303	11/16	25.406	17.8678
3/8	0.11045	1.17810	3/4	5.9396	8.63938	3/4	25.967	18.0642
13/32	0.12962	1.27627	13/16	6.2126	8.83573	13/16	26.535	18.2605
7/16	0.15033	1.37445	7/8	6.4918	9.03208	7/8	27.109	18.4569
15/32	0.17257	1.47262	15/16	6.7771	9.22843	15/16	27.688	18.6532
1/2	0.19635	1.57080	3	7.0686	9.42478	6	28.274	18.8496
17/32	0.22166	1.66897	1/16	7.3662	9.62113	1/8	29.465	19.2423
9/16	0.24850	1.76715	1/8	7.6699	9.81748	1/4	30.680	19.6350
19/32	0.27688	1.86532	3/16	7.9798	10.0138	3/8	31.919	20.0277
5/8	0.30680	1.96350	1/4	8.2958	10.2102	1/2	33.183	20.4204
21/32	0.33824	2.06167	5/16	8.6179	10.4065	5/8	34.472	20.8131
11/16	0.37122	2.15984	3/8	8.9462	10.6029	3/4	35.785	21.2058
23/32	0.40574	2.25802	7/16	9.2806	10.7992	7/8	37.122	21.5984
3/4	0.44179	2.35619	1/2	9.6211	10.9956	7	38.485	21.9911
25/32	0.47937	2.45437	9/16	9.9678	11.1919	1/8	39.871	22.3838
13/16	0.51849	2.55254	5/8	10.321	11.3883	1/4	41.282	22.7765
27/32	0.55914	2.65072	11/16	10.680	11.5846	3/8	42.718	23.1692
7/8	0.60132	2.74889	3/4	11.045	11.7810	1/2	44.179	23.5619
29/32	0.64504	2.84707	13/16	11.416	11.9773	5/8	45.664	23.9546
15/16	0.69029	2.94524	7/8	11.793	12.1737	3/4	47.173	24.3473
31/32	0.73708	3.04342	15/16	12.177	12.3700	7/8	48.707	24.7400
1	0.78540	3.14159	4	12.566	12.5664	8	50.265	25.1327
1/16	0.88664	3.33794	1/16	12.962	12.7627	1/8	51.849	25.5224
1/8	0.99402	3.53429	1/8	13.364	12.9591	1/4	53.456	25.9181
3/16	1.1075	3.73064	3/16	13.772	13.1554	3/8	55.088	26.3108
1/4	1.2272	3.92699	1/4	14.186	13.3518	1/2	56.745	26.7035
5/16	1.3530	4.12334	5/16	14.607	13.5481	5/8	58.426	27.0962
3/8	1.4849	4.31969	3/8	15.033	13.7445	3/4	60.132	27.4889
7/16	1.6230	4.51604	7/16	15.466	13.9408	7/8	61.862	27.8816
1/2	1.7671	4.71239	1/2	15.904	14.1372	9	63.617	28.2743
9/16	1.9175	4.90874	9/16	16.349	14.3335	1/8	65.397	28.6670
5/8	2.0739	5.10509	5/8	16.800	14.5299	1/4	67.201	29.0597
11/16	2.2365	5.30144	11/16	17.257	14.7262	3/8	69.029	29.4524
3/4	2.4053	5.49779	3/4	17.721	14.9226	1/2	70.882	29.8451
13/16	2.5802	5.69414	13/16	18.190	15.1189	5/8	72.760	30.2378
7/8	2.7612	5.89049	7/8	18.665	15.3153	3/4	74.662	30.6305
15/16	2.9483	6.08684	15/16	19.147	15.5116	7/8	76.589	31.0232
						10	78.540	31.4159

CONVENTIONAL SECTION LINES

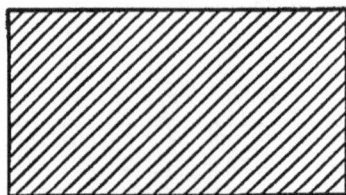

CAST IRON

WROUGHT IRON

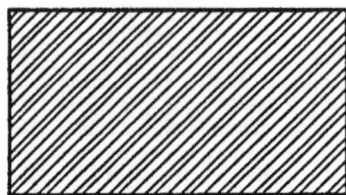

MALLEABLE IRON

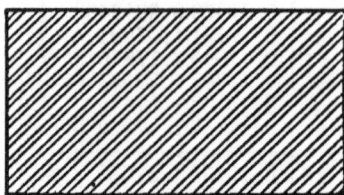

CAST STEEL

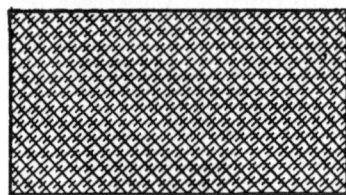

COPPER

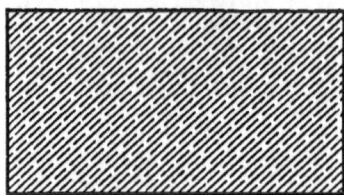

BRASS OR BRONZE

BABBITT

VULCANITE

GLASS

WATER